The Essence of Software

The Essence of Software

Why Concepts Matter for Great Design

Daniel Jackson

PRINCETON UNIVERSITY PRESS · PRINCETON & OXFORD

Published by Princeton University Press
41 William Street, Princeton, New Jersey 08540
6 Oxford Street, Woodstock, Oxfordshire OX20 1TR
press.princeton.edu

All Rights Reserved
ISBN 978-0-691-22538-8
ISBN (e-book) 978-0-691-23054-2

British Library Cataloging-in-Publication Data is available
Editorial: Hallie Stebbins and Kristen Hop
Production Editorial: Jenny Wolkowicki
Jacket design: Emily Weigel
Production: Danielle Amatucci
Publicity: Sara Henning-Stout and Kate Farquhar-Thomson
Copyeditor: Bhisham Bherwani
Interior design: Daniel Jackson
Set in Adobe Arno Pro and Stone Magma using Adobe InDesign
Printed on acid-free paper. ∞
Printed in the United States of America
10 9 8 7 6 5 4 3 2 1

to my parents

Contents

How to Read This Book

A micromaniac is someone obsessed with reducing things to their
smallest possible form. This word, by the way, is not in the dictionary.
—*Edouard de Pomiane,* French Cooking in Ten Minutes

Concept design is a simple idea that you'll be able to apply in your own work—using and designing software—without having to master any complex technicalities. Many of the concepts I'll use as examples will be recognizable old friends. So I'll take it as a compliment if your conclusion, after reading this book, is that concepts are a natural and even obvious way to think about software, and that you learned nothing more than a systematic framework for an intuitive idea.

But even if the underlying theme of this book resonates and seems familiar, I suspect that for many readers this new way of thinking about software will be disorienting, at least initially. Although software designers have talked for decades about conceptual models and their importance, concepts have never been placed at the center of software design. What would design look like if every software app or system were described in terms of concepts? What exactly would those concepts be? How would they be structured? And how would they be composed together to form the product as a whole?

To answer these questions as best I can, this book is longer than I would have liked. To mitigate that, I have organized it so that different readers can take different journeys through it. Some will want to reach a pragmatic destination as quickly and expeditiously as possible; others, wanting a deeper understanding, may prefer to follow me on some detours away from the main path. This little guide should help you plan your route.

Intended Audience

In short, this book is aimed at anyone interested in software, in design, or in usability. You might be a programmer, a software architect, or a user interaction designer; a consultant, an analyst, a program manager, or a marketing strategist; a computer science student, teacher, or researcher; or maybe just one of those people who (like me) enjoys thinking about why things are designed in certain ways—and why some designs succeed so gloriously and others fail so spectacularly.

No knowledge of computer science or programming is assumed, and although many of the principles in the book can be expressed more precisely in logic, no mathematical background is required. In order to appeal to as broad an audience as possible, I've drawn examples from a wide variety of popular apps, from word processors to social media platforms. So each reader will likely encounter examples that are easy to follow and others that require more effort. A byproduct of reading the book, I hope, will be a more solid grasp (and thus greater mastery) of apps that you use but don't yet fully understand.

Goals of This Book

This book has three related goals. The first is to present some straightforward techniques that software creators can apply immediately to improve the quality of their designs. By helping you identify and disentangle the essential concepts, articulate them, and make them clear and robust, the book will enable you to design software better—and thus design better software—whatever phase of design you work in, from the earliest phases of strategic design (in which products are imagined and shaped) to the latest phases (in which every detail of interaction with the user is settled).

The second goal is to provide a fresh take on software, so you can view a software product not just as a mass of intertwined functions, but as a systematic composition of concepts, some classic and well understood, and others novel and more idiosyncratic. With this new perspective, designers can focus their work more effectively, and users can understand software with greater clarity, empowering them to exploit their software to its fullest potential.

My third and final goal is broader and perhaps easier. It is to convince the community of researchers and practitioners who work on the development of software applications and services that the design of software is an exciting and intellectually substantive discipline.

Interest in software design—especially when focused on user-facing aspects—has waned over the last few decades, even as recognition of its importance has grown. In part this is due to the misconception that there is nothing inherent to the design of software that makes it more or less usable, and that any such judgments are subjective (or better addressed as psychological or social questions, with the focus more on the user than the software itself).

The rise of empiricism in software practice—while motivated by the apt recognition that even the best designs have flaws that only user testing will reveal—has also, in my view, dulled our enthusiasm for design, as many have come to doubt the value of design expertise. But mostly, I believe, we have suffered from a lack of respectability and intellectual confidence, since our ideas about what makes software usable have more often been expressed as tentative rules of thumb rather than principles grounded in a rich theory. I hope to show, in this book, that such principles and theory do indeed exist, and to encourage others to pursue their development and refinement.

Choosing Your Path

You can take different paths through the book, depending on your goals. To do so, it will help you to know how the book is organized and what each part contains.

Part I contains three motivational chapters. The first might have served as a preface: it explains why I came to write this book, and why the problems that I was interested in hadn't already been addressed in other fields (such as human-computer interaction, software engineering and design thinking). In the second chapter, we see our first examples of concepts, and the impact that they have on usability, and I explain concept design as the top of a hierarchy of user-experience design levels. The third chapter outlines the many roles that concepts have, from being product differentiators to being the linchpin of digital transformations.

Part II is the heart of the book. Its first chapter tells you exactly what a concept is, and how it can be structured. The second explains the fundamental idea of a concept's purpose as a motivation and a yardstick. The third shows how an app or system can be understood as a composition of concepts, combined using a simple but powerful synchronization mechanism; it explains how over- or under-synchronization can damage usability; and, more subtly, how some features that we traditionally view as complex and indivisible can instead be understood as synergistic fusions of distinct concepts. The fourth shows how mapping concepts to a user interface is not always as straightforward as you might imagine, and that sometimes the problem with a design lies not with the concepts per se but in their realization as buttons and displays. The last chapter in this part introduces a way to think about software structure at a very high level as a collection of concepts that are mutually dependent on one another—not that any concept relies on another concept for its correct working, but rather that only certain combinations of concepts make sense in the context of an app.

Part III introduces three key principles of concept design each in its own chapter: that concepts should be specific (one-to-one with purposes); that they should be familiar; and that the integrity of a concept should not be violated in composition (resulting in behaviors that, when viewed through the lens of an individual concept, do not satisfy that concept's specification).

The body of the book closes with a list of provocative questions addressed to readers in different roles. You might use these as a summary of the lessons of the book; or as a checklist in your subsequent practice; or even read them first as a quick overview of what the book offers.

If you want to jump in at the deep end, you could start with Part II; the motivations of Part I could be read as a conclusion, summarizing the ways in which the ideas you've learned can be applied. Each chapter ends with a summary of the main lessons and a list of practices that you can apply immediately.

Explorations & Digressions

Almost half the book is a collection of endnotes. I wrote the book like this because I wanted to make the body of the book as brief as possible, while also carefully justifying my approach and explaining its connection to existing design theories. So the main part of the book contains no discussion of related work (not even a single citation), ignores many subtle points, and leaves out many ideas I have about design more generally.

The endnotes make up for these omissions. There, I not only cite related work but attempt to put it all in context and explain its significance. I explain in much more detail the distinguishing characteristics of concept design, and I present examples that require more background (or persistence) to comprehend. When I have not managed to resist the temptation to fulminate (against rampant empiricism, or a myopic focus on defect elimination, for example), I have at least relegated my diatribes to these notes.

The notes are cited at appropriate points in the main text with superscript numerals; the first appears a few sentences from here. But to spare you the annoyance of flipping back and forth, I have grouped them into free standing sections with their own titles, so they can be read independently, and you can just dive in and read them randomly at your leisure.

Multiple Indexes

Rather than just one index, this book has four separate ones: an index of the applications from which examples are drawn; an index of concepts; an index of names; and a general index of topics. The entries under *concept* in the index of topics might be particularly useful, as they highlight the key qualities of concepts and point to various mini essays in the endnotes.

Warning: Micromaniac at Work

De Pomiane, author of the inimitable *French Cooking in Ten Minutes*,[1] confesses in his introduction to being a micromaniac. I readily admit to the same pathology. I don't want to hear that a design failed or succeeded for a myriad of amorphous reasons. Even if that were sometimes true, what use would it be? I want to get to the essence, to put my finger on the one essential spot, on the crucial design decision that launched the product to dizzying success or sunk the entire enterprise.

I am not naive, and I know that being cognizant of multiple factors in design—especially when analyzing the causes of accidents—is wise and sensible. But it's just not a valuable way to draw lessons from prior experiences. To do that, I believe we all need to be micromaniacs: to focus on the tiniest of details in the search for an elusive but potent explanation whose generalization offers a lasting and widely applicable lesson. So be warned: the devil is in the details—and the angels are too.[2]

PART I
MOTIVATIONS

1

Why I Wrote This Book

As an undergraduate in physics, I'd been entranced by the idea that the world could be captured by simple equations like $F = ma$. When I became a programmer, and later a computer science researcher, I gravitated towards the field of formal methods, because it promised to do something similar for software: to express its very essence in a succinct logic.

A Passion for Design

My main research contribution in the 30 years since my PhD has been Alloy,[3] a language for describing software designs and analyzing them automatically. It's been an exciting and satisfying journey for me, but I came to realize over time that the essence of software doesn't lie in any logic or analysis. What really fascinated me wasn't the question that consumed most formal methods researchers—namely how to check that a program's behavior conforms exactly to its specification—but rather the question of *design*.[4]

I mean "design" here in the same sense that the word is used in other design disciplines: the shaping of some artifact to meet a human need. Design, as the architect Christopher Alexander put it, is about creating a *form* to fit a *context*. For software, that means determining what the behavior of the software should be: what controls it will offer, and what responses it will provide in return. These questions have no right or wrong answers, only better or worse ones.[5]

I wanted to know why some software products seem so natural and elegant, react predictably once you master the basics, and let you combine their features in powerful ways. And to pinpoint why other products just seem wrong: cluttered with needless complexity, and behaving in unexpected and inconsistent ways. Surely, I thought, there must be some essential principles, some theory of software design, that could explain all of this. It would not only explain why some software products are good and some are bad, but it would help you fix the problems and avoid them in the first place.

Design in Computer Science and Other Fields

I started to look around. Within my own subfield (formal methods, software engineering and programming languages), such a theory exists for what you might call "internal design"—namely the design of the structure of the code. Programmers have a rich language of design, and well-established criteria for what distinguishes good designs from bad ones. But no such language or criteria exist for software design in the user-facing sense, namely design that determines how software is experienced as a form in context.[6]

Internal code design is very important and influences primarily what software engineers call "maintainability," which means how easy (or hard) the code is to change over time as needs evolve. It also influences performance and reliability. But the key decisions that determine whether a software application or system is useful and fulfills its users' needs lie elsewhere, in the kind of software design in which the functionality and the patterns of interaction with the user are shaped.

These big questions were at one time more central in computer science. In the field of software engineering, they came up in workshops on software design, specification and requirements; in the field of human-computer interaction, they permeated early work on graphical user interfaces and computational models of user behavior.[7]

But as time passed, they became less fashionable, and they faded away. Research in software engineering narrowed, and eliminating defects—whether by testing or more sophisticated means such as program verification—became synonymous with software quality.[8] But you can't get there from here: if your software has the wrong design, there's no amount of defect elimination that will fix it, short of going back to the very start and fixing the design itself.[9]

Research in human-computer interaction (HCI) shifted to novel interaction technologies, to tools and frameworks, to niche domains, and to other disciplines (such as ethnography and sociology). Both software engineering and HCI embraced empiricism enthusiastically, largely in the misguided hope that this would bring respectability. Instead, the demand for concrete measures of success seems to have led researchers towards less ambitious projects that admit easier evaluation, and has stymied progress on bigger and more important questions.[10]

Puzzlingly, even as interest in design seems to have waned, talk of "design" is everywhere. This is not in fact a contradiction. The talk, almost exclusively, is about the *process* of design, whether in the context of "design thinking" (a compelling packaging of iterative design processes), or of "agile" software development. These processes are undoubtedly valuable (so long as they are applied judiciously and not as panaceas), but they are for the most part content-free. I mean that not to disparage but to describe. Design thinking, for example, might tell you to develop your solution hand in hand with your understanding of the problem, or to engage in alternating phases of brainstorming ("divergence") and reduction ("convergence"). But no design thinking book that I have read talks in depth about any particular designs and how the process sheds light on them. The very domain-independence of design thinking may be the key to its widespread appeal and applicability—but also the reason it has little to say about deeper challenges of design in a particular domain such as software.[11]

Clarity & Simplicity in Design

When I began the Alloy project, with the goal of creating a design language that was amenable to automatic analysis, I was critical of existing modeling and specification languages whose lack of tool support rendered them "write-only." This snide dismissal was not entirely unwarranted. After all, why would you go to the trouble of constructing an elaborate design model if you couldn't then do anything with it? I argued, in particular, that the designer's effort should be rewarded immediately with "push-button automation" that would instantly give you feedback in the form of surprising scenarios that would challenge you to think more deeply about your design.[12]

I don't think I was wrong, and Alloy's automation did indeed change the experience of design modeling. But I had underestimated the value of writing down a design. In fact, it was a not very well guarded secret amongst formal methods researchers (who were eager to demonstrate the efficacy of their tools by finding flaws in existing designs) that a high proportion of the flaws were detected *before* the tools were even run! Just transcribing the design into logic was enough to reveal serious problems. The software engineering researcher Michael Jackson credits not the logic per se but the very difficulty of using it, and once mischievously suggested that the quality of software systems might be improved if designers were simply required to record their designs in Latin.

Clarity is good not only for finding design flaws after the fact. It is also the key to good design in the first place. In teaching programming and software engineering over the last thirty years, I've become increasingly convinced that the determinant of success when you're developing software isn't whether you use the latest programming languages and tools, or the management process you follow (agile or otherwise), or even how you structure the code. It's simply whether you know what you are trying to do. If your goals are clear, and your design is clear—and it's clear how your design meets the goals—your code will tend to be clear too. And if something isn't working, it will be clear how to fix it.[13]

It is this clarity that distinguishes great software from the rest. When the Apple Macintosh came out in 1984, people could see immediately how to use folders to organize their files; the complexities of previous operating systems (such as Unix, which made even the command to move files between folders complicated) seemed to have evaporated.

But what exactly is this clarity, and how is it achieved? As early as the 1960s, the central role of "conceptual models" has been recognized. The challenge was not merely to *convey* the software's conceptual model to the user so that her internal version ("mental model") was aligned with the programmers', but to treat it as a subject of design in its own right. With the right conceptual model, the software would be easy to understand and thus easy to use. This was a great idea, but nobody seems to have pursued it, and so until now "concepts" have remained a vague, if inspiring, notion.[14]

How This Project Came About

Convinced that conceptual models were indeed the essence of software, I started about eight years ago trying to figure out what they might be. I wanted to give them concrete expression, so that I could point to some software's conceptual model, compare it to others (and to the mental models of users), and have an explicit focus for design discussions.

That didn't seem so hard. After all, a plausible first cut at a conceptual model might be just a description of the software's behavior, made suitably abstract to remove incidental and "non-conceptual" aspects (such as the details of the physical user interface). What proved much harder was finding appropriate structure in the model. I had an inkling that a conceptual model should be made up of concepts, but I didn't know what a concept was.

In a social media app such as Facebook, for example, it seemed to me that there should be a concept associated with liking things. This concept surely wasn't a function or action (such as the behavior bound to the button you click to like a post); there are too many of those, and they only tell part of the story. It also surely wasn't an object or entity (such as the "like" itself that your action produced), since at the very least the concept seemed to be about the *relationship* between things and their likes. It also seemed essential to me that the concept of liking was not associated with any particular kind of thing: you could like posts, comments, pages, and so on. The concept, in programming lingo, is "generic" or "polymorphic."

This Book: Opening a Conversation

This book is the result of my explorations to date. Driven by dozens of design issues in widely used applications, I've evolved a new approach to software design, refining and testing it along the way. A happy aspect of this project has been that every app failure or frustration had a silver lining: a chance to extend my repertoire of examples. It has also given me greater sympathy and respect for the designers when my analysis revealed the full complexity of the problem they faced.

Of course, the problem of software design is not solved. But as my friend Kirsten Olson wisely advised me: a book should aim to start a conversation, not to end one. In the course of giving many talks about this project, I've been thrilled to discover that it seems to resonate with audiences more than any of my previous ones. I suspect this is because software design is something we all want to talk about, but we have not known how to have that conversation.

So to you, my readers—fellow researchers, designers and users—I present this book as my opening gambit in what I hope to be a fruitful and enjoyable conversation.

2

Discovering Concepts

A software product—from the smallest app that runs on your phone to the largest enterprise system—is made of *concepts*, each a self-contained unit of functionality. Even though concepts work in tandem for a larger purpose, they can be understood independently of one another. If an app is like a chemical mixture, concepts are like molecules: although bound together, their properties and behavior are similar wherever they are found.

You're already familiar with many concepts, and know how to interact with them. You know how to place a phone *call* or make a restaurant *reservation*, how to *upvote* a comment in a social media forum and how to organize files in a *folder*. An app whose concepts are familiar and well designed is likely to be easy to use, so long as its concepts are represented faithfully in the user interface and programmed correctly. In contrast, an app whose concepts are complex or clunky is unlikely to work well, no matter how fancy the presentation or clever the algorithms.

Since concepts have no visible form, they're rather abstract, and this is perhaps why they haven't been a focus of attention until now. I hope to persuade you, in the course of this book, that by thinking in terms of concepts, and by "seeing through" user interfaces to the concepts that lie behind them, you will be able to understand software more deeply—to use it more effectively, to design it better, to diagnose flaws more precisely, and to envision new products with greater focus and confidence.

We don't generally appreciate how something works until it breaks. You may think that your water heater just magically produces a constant stream of hot water. But then at some point someone in your household takes one shower too many, and your shower is cold. That's when you might learn that your water heater has a *storage tank* with limited capacity.

Likewise, to learn about concepts, we need to see what happens when they go wrong. Much of this book, therefore, will involve examples of concepts that fail in seemingly unlikely scenarios, or that turn out to be much harder to understand than you'd expect them to be. In this chapter, we'll see our first examples of concepts, and how they can explain some unexpected (and surprisingly complicated) behaviors.

But don't be put off, or draw the conclusion that the *idea* of concepts is itself obscure and complicated. On the contrary, the idea is straightforward, and adopting it will help you to design software that is simpler and more powerful than much of the software we use today.

A First Example: Baffling Backups

To protect my work from corrupted disks and accidental deletion, I use a terrific backup utility called Backblaze, which copies my files to the cloud, and lets me restore old versions if I need to. It runs invisibly and continuously in the background, keeping an eye on every file in my computer, copying it to the cloud if it changes.

Recently, I edited a video and wanted to make sure the new version had been backed up before I deleted the old one to save space. I checked the backup status, and it said "You are backed up as of: Today, 1:05 PM." Since I had created the new video *before* 1:05 PM, I assumed it had been backed up. Just to be sure, I tried to restore it from the cloud. But it wasn't there.

I contacted tech support, and they explained to me that files aren't exactly backed up continuously. There's a periodic scan that compiles a list of new or modified files; when the next backup runs, only files on that list are uploaded. So any changes made between the scan and the backup fall between the cracks until they're discovered in the next scan.

I could force a rescan, they told me, by clicking the "Backup Now" button while holding down the option key. I followed this advice, and waited for the scan and subsequent backup to complete. Now, surely, my new video would show up on the restore list! But no such luck. At this point, I was totally confused, and asked for more help. It turned out that my video *had* been uploaded, but only to a special "staging" area, from which files are moved to the restore area every few hours.

My problem was that I misunderstood the key *backup* concept of Backblaze. I had imagined that files were uploaded continuously, and moved directly to the restore area (Figure 2.1, left). In fact, only the files on the list produced by the last

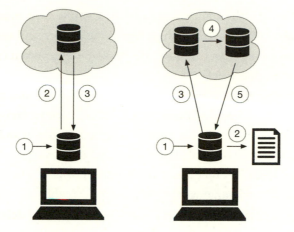

FIG. 2.1 *Backblaze's backup concept. On the left, what I assumed: (1) I make a change to a file; (2) when the backup runs, the file is copied to the cloud; (3) I can then restore it. On the right, what actually happens: (1) I make a change to a file; (2) a scan runs and adds the file to a list of files for backup; (3) the backup runs, copying to the cloud only those files that were added in the last scan; (4) periodically, backed-up files are moved to a cloud location (5) from where they can be restored.*

scan are uploaded, and even then remain unavailable until they have been transferred sometime later from the upload destination to the restore area (Figure 2.1, right).

This is a small example but it illustrates my key point. I'm not taking a stand on whether the design of Backblaze is flawed or not; I suspect it could be improved though (see Chapter 8 for a suggestion). Certainly, had I taken the backup message at face value and not known about the scan, I might have lost some crucial files.

What I *am* claiming is that any discussion of this design must revolve around the fundamental concepts, in this case the *backup* concept, and an assessment of whether the behavioral pattern that it embodies is fit for purpose. The user interface matters too, but only to the extent that it serves the app's concepts by representing them to the user. If we want to make software more usable, concepts are where we must start.

Dropbox Delusions

A friend of mine was running out of space on her laptop. So she cleverly sorted the files by size, and looked down the list to see if there were any large and unfamiliar files that she could get rid of. She identified a dozen or so such files, and went ahead and deleted them. A few minutes later she got a panicked call from her boss asking what had happened to some large files containing data for an important work project.

What went wrong? To answer this, we need to understand some key concepts of Dropbox, a popular file-sharing utility. Dropbox allows multiple users to view a shared collection of files and folders, and to update them collaboratively. To maintain this illusion, Dropbox propagates changes made by one user to the versions seen by other users. The question will be: what kinds of changes are propagated? And under what conditions?

Ava is a party planner who uses Dropbox to coordinate with her customers. She's planning a party for Bella, so she creates a folder with the name *Bella Party* and shares it with Bella (Figure 2.2). Whatever Ava puts into the folder, Bella can now see. In fact, the sharing is symmetrical; whatever Bella puts in, Ava can see too, and whatever changes one of them makes, the other sees those very same changes. So it's as if there were just one copy of a folder that Ava and Bella can work on together.

Actually, it's not quite that simple, because not *all* changes that one of them makes will be seen by the other. Perhaps Bella doesn't want the folder to be called *Bella Party*—after all, it's her party! So she gives the folder the new name *My Party*. The question is: what does Ava now see? Does the name change for her too?

There are only two possibilities. Either Bella's action will change the name Ava sees too, in which case there is just one shared file name, or it will not, in which case there are two names for the same folder, one that Ava uses and one that Bella uses.

So which happens? It turns out that *both* outcomes are possible, depending on how the folder was shared. In this case, in which Ava has shared the folder *explicitly* with Bella, Bella's renaming will only be seen by Bella, and Ava will *not* see the change. But suppose Ava creates another folder inside *Bella Party* called *Bella Plan* (Figure 2.3 top). This second folder is now shared *implicitly* (by virtue of its containing folder being shared). Now if Bella renames *Bella Plan* to *My Plan*, say, then Ava *will* see the change.

You might imagine that this variability of behaviors is accidental, that it's the result of some arbitrary choices that arose during Dropbox's evolution. Or you

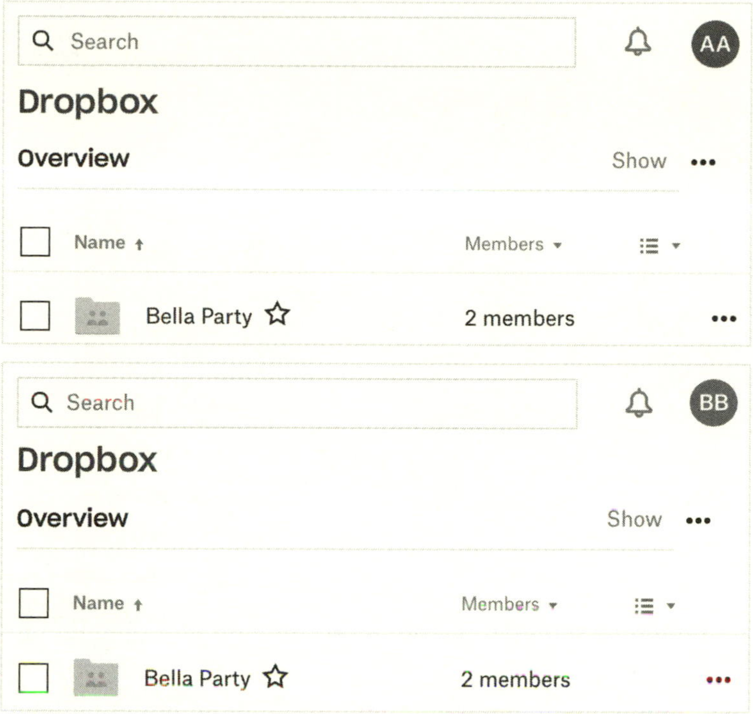

FIG. 2.2 *Sharing a folder in Dropbox. Ava (AA) has shared the folder named Bella Party with Bella (BB). If Bella now changes the name of the folder, will Ava see the change?*

might think it's evidence of a bug. In fact, neither of these is true. This apparent design oddity is a direct consequence of a fundamental aspect of Dropbox's design.

Before I explain exactly what's going on, let's consider one more question. What happens if Bella *deletes* a folder? Will Ava's copy be deleted too? Again, it depends on the context. If Bella deletes *Bella Party*, her copy alone will go away; but if she deletes *Bella Plan*, Ava will lose it too. Dropbox does give different messages in the two cases (Figure 2.3), one of which explains more fully what will happen. But, strangely, the additional information is given in the first case, not in the second case when the deletion results in permanent loss of the file.

Now we have an explanation for my friend's experience. Her boss, wanting to share just one file with her, had shared the entire folder instead. When my friend removed the files she wasn't using, she was deleting them from the shared folder—and thus removing them for everyone, her boss included.

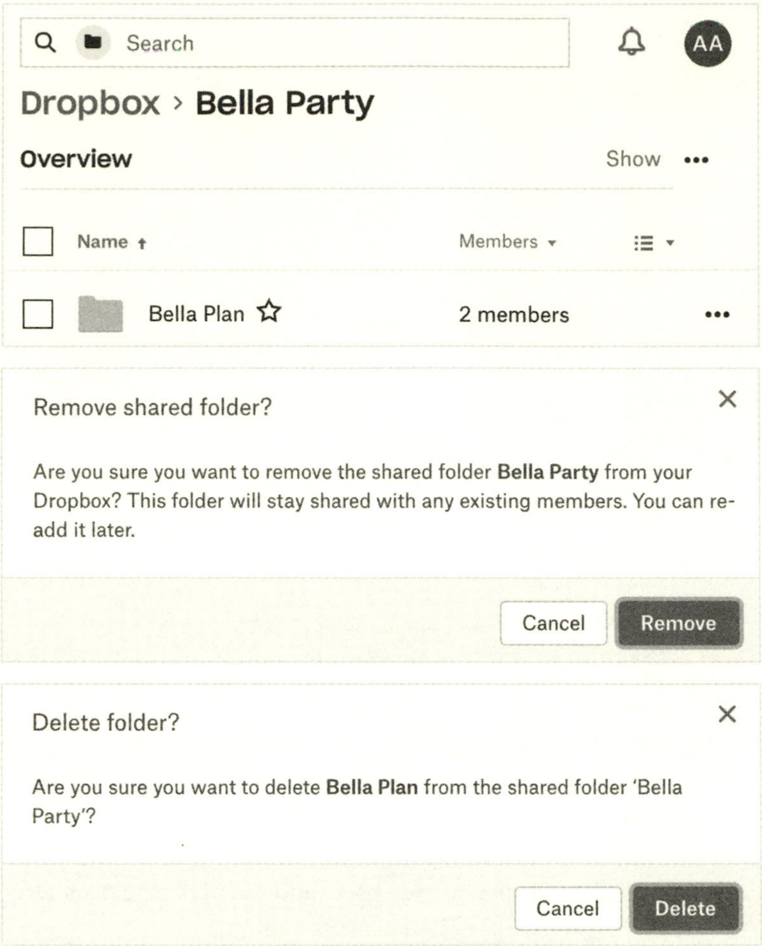

FIG. 2.3 *Dropbox's folder deletion messages. The folder Bella Party has been shared (top). If that folder is deleted, the message (middle) informs you that the deletion will not be propagated to other users. If the folder Bella Plan contained within it is deleted, a different message (bottom) appears, surprisingly not warning that other users will lose the folder too.*

Explaining Dropbox

To see what's going on in these sharing scenarios, it helps first to articulate what our expectations might have been. A simple and familiar design for names would treat them as if they were sticky labels attached to physical objects—like a cat collar, or a license plate—with at most one label per object (Figure 2.4, left). We might call this approach "name as metadata," and it would be an in-

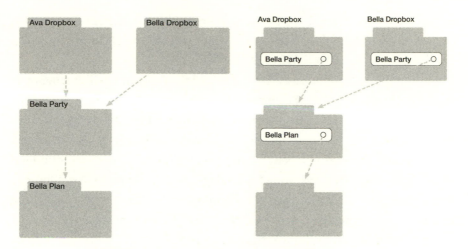

FIG. 2.4 *Two possible concepts for folders in Dropbox: in the metadata concept (left), names are labels attached to folders; in the unix folder concept (right), names belong to entries within the parent folder.*

stance of a more general *metadata* concept in which data that describes an object—such as the title or caption of a photo—can be attached to it.

With respect to deletion, the simplest design would be to make a file or folder disappear when it's deleted. We might call this (using a technical term) the "deletion as poof" approach to deletion: you click delete, and "poof!"—it's gone. The underlying concept here—that a pool of items can be stored, with actions to add and remove items from the pool—is so basic and familiar it has no name. In this design, we'd expect a separate *sharing* concept, with an *unshare* action so that you can remove a file or folder that someone else shared with you and free up the space in your own account without deleting their copy.

Both of these understandings—that names are metadata and deletion simply removes items from a pool—are wrong (at least for Dropbox). The concepts behind these understandings are themselves fine; they're just not the concepts that Dropbox uses. If you hold the wrong conceptual model of a software app, you might get away with it for a while. We've seen that in some scenarios these explanations would work successfully. But in other scenarios, they'll fail, perhaps with disastrous consequence.

The actual concepts that Dropbox uses are very different (Figure 2.4, right). When an item sits in a folder, the name of that item belongs *not* to the item itself but rather to the folder containing it. Think of a folder as being a collection of tags, each containing the name of an item (a file or folder) and a link to it. This concept, which I'll call *unix folder*, was not invented by Dropbox, but, as its name suggests, was borrowed from Unix.[15]

Look at the diagram in Figure 2.4 (right). Each of Ava and Bella has her own, top-level Dropbox folder, and these two folders have *separate* entries for the single, shared folder called *Bella Party*. When Bella renames *Bella Party*, this alters the entry in her own Dropbox folder, and the entry in Ava's folder is unchanged.

In contrast, there is only a single entry holding the name of the second-level shared folder, *Bella Plan*, belonging to the single, shared parent folder called *Bella Party*. Since there is only *one* entry for the folder—the same entry seen by both Ava and Bella—when Bella renames the folder, she is changing that one entry in their shared folder, so Ava sees the change too.

Using this same *unix folder* concept, we can now explain the deletion behaviors. Deletion doesn't remove the folder per se; it removes its entry. So if Bella deletes the folder *Bella Party*, she removes the entry from her own folder, and Ava's view is unchanged. But if Bella deletes *Bella Plan*, she removes the entry from the shared folder, and the deleted folder is now inaccessible to Ava too.

What Kind of Flaw is This?

At this point, you might be saying to yourself: Well, this is all obvious. I knew Dropbox behaved like this and I'm not in the least bit surprised. There's nothing wrong with Dropbox, and someone who doesn't understand it shouldn't be using it. But if you think this, I'm pretty sure you'd be in the minority of readers. We presented this scenario to MIT computer science students and found that many of them, even those who used Dropbox regularly, were confused.[16]

Even if you understood all these subtleties, I'd argue that there's still a problem. The distinction between the two cases—whether the folder that is the subject of the action is shared at the top level, or belongs to another folder that is itself shared—isn't readily discernible in the user interface, so it's a constant annoyance having to figure out which situation you're in.

Moreover, it doesn't seem reasonable that this rather arbitrary distinction should determine the behavior. Why should I be able to give my own name only to the top-level folder? Why can't I give private names to all the folders shared with me? Or conversely, if renaming folders for both of us is part of our shared work, why can I only do it for some folders and not others?

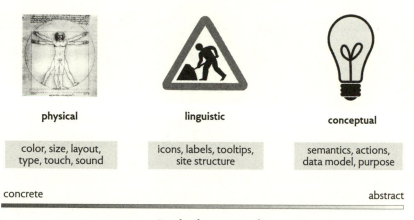

physical	linguistic	conceptual
color, size, layout, type, touch, sound	icons, labels, tooltips, site structure	semantics, actions, data model, purpose

concrete   abstract

FIG. 2.5 *Levels of interaction design.*

Assuming then, that these scenarios are indeed evidence of a flaw in Dropbox, we can ask: what kind of flaw is it? It's certainly not a bug; Dropbox has behaved like this for years. We might wonder if it's a flaw in the user interface. That seems implausible too. It would be possible, of course, for Dropbox to give more informative messages when a change you make affects other users. But this might just be perceived as additional complexity, and experience suggests that users ignore warning messages if they come up too often.[17]

The real problem runs deeper. It's in the very essence of how files and folders are named, and how those names are related to the containment relationship between folders and their contents. This is what I call a *conceptual* design issue. The flaw is that the Dropbox developer has certain concepts in mind that have been faithfully implemented. But those concepts, at the very least, are not consistent with the concepts in most users' minds. And, at worst, these concepts are not a good match for the users' purposes.[18]

Levels of Design

To put conceptual design in perspective, it helps to break software design into levels, as shown in Figure 2.5. This classification is my own, but it is similar to schemes previously proposed.[19]

The first level of design, the *physical level*, is about the physical qualities of the artifact. Even software whose interface is no more than a touch-sensitive piece of glass has such qualities, limited though they might be.[20] At this level, the designer must take into account physical capabilities of human beings. It's

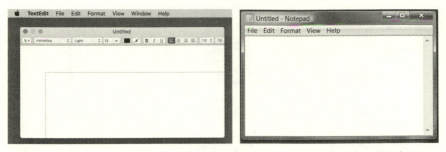

FIG. 2.6 *A design issue at the physical level, and a classic example of applying Fitts's Law. Which menu placement allows for more convenient access: the* macos *placement (on the left) in which an application's menu bar always appears at the top of the desktop, or the Windows placement (on the right) in which the menu bar is part of the application window?*

where accessibility concerns arise, as the designer considers how a visually impaired, or color-blind, or deaf user might interact.

Common human characteristics dictate certain design principles. For example, the fact that our limited visual sampling rate results in *perceptual fusion*, making it hard to distinguish events that occur within about 30 ms of each other, suggests that 30 frames/second is enough for a movie to look smooth. It also tells us that system reactions that take much longer than 30 ms will be perceived as delays by the user, and should be avoided, or given progress bars, and if very much longer, an opportunity to abort. Likewise, Fitts's Law predicts the time it takes for a user to move a pointing device to a target, and explains why the menu bar should be positioned at the top of the screen, as in the Macintosh desktop, and not in the application window, as in Windows (Figure 2.6).[21]

The second level of design is the *linguistic level*. This level concerns the use of language for conveying the behavior offered by the software, to help the user navigate the software, understand what actions are available and what impact they will have, what has happened already, and so on. While design at the physical level must respect diversity amongst the physical characteristics of its users, design at this level must respect differences of culture and language.

Obviously, the button labels and tooltips on an app will vary depending on whether it's intended for English or Italian speakers. (I remember as a small child on holiday in Italy learning the hard way that the faucet marked *calda* is not the cold one.) The designer must be aware of cultural differences too. In Europe, a road sign comprising a red circle with a white interior means that no vehicular traffic is permitted at all; most American drivers would not be able to

FIG. 2.7 *A design issue at the linguistic level: inconsistent interpretation of icons in Google apps and how new icons fixed the problem. On the left, the original icons for opening apps and switching to grid view; on the right the new icons for the same actions, now distinguished.*

interpret this (and might expect the prohibition to be expressed instead with a red diagonal bar).[22]

When user interface designers talk about the need for consistency, they're usually referring to the use of language at this level. Consistency includes ensuring that the same words are used in the same way throughout an interface—for example, that containers for files aren't called "folders" in one place and "directories" in another—and that icons are used systematically. Figure 2.7 shows how Google violated this principle by using two near-identical icons for different functions (but fixed the problem a few years later). Both depicted an array of black squares, but one opened a menu of Google apps, and the other switched to a view in which files appear in a grid.

The third, and highest, level of design is the *conceptual level*. It's concerned with the behavior that underlies the design: the actions that are performed by the user (and by the software itself) and the consequences that they have on underlying structures. In contrast to the linguistic level, the conceptual level isn't about communication or culture, even though (as we'll see in Chapter 10) prior knowledge of a concept makes it easier to learn and use.

In programming, there is a familiar and important distinction between *abstraction* and *representation*. An abstraction captures the essence of a programming idea, and may be expressed as a specification of the observable behavior. A representation is the realization of that essence in code.

In the same way, a user interaction has an abstraction and a representation. The abstraction is the concept—the essence of the structure and behavior—and is the subject of design at the conceptual level. The representation is the realization of the concept in a user interface, with all its physical and linguistic details, and is the subject of design at the lower levels.

Just as a single programming abstraction can have different representations, so a concept can be realized in different user interfaces. And just as programmers think first about the abstraction and only second about the representation, so designers think at the conceptual level before the lower levels. Until now, designers have not had a way to express conceptual design ideas without representing them in a concrete user interface. The goal of this book is to show that such ideas can be expressed directly, prior to and independently of representation choices.

Mental Models and Concept Design

In most software apps, the difficulties that users have are rarely due to an app having too few features or too many. The more common problem is that users can't make effective use of the functionality that actually exists. This may happen because users don't actively seek out features, and assume that they're not there.[23]

More often, though, users are aware of features but are just unable to use them successfully. The most common reason for this is that the user has a mental model that is incorrect—that is, incompatible with the mental model of the designer and implementer of the software. Although research has shown repeatedly (and unsurprisingly) that users often have vague, incomplete and even inconsistent models of the devices they use, they do develop conceptions that are similar, at least in form if not in every detail, to the conceptions held in the minds of the designers.[24]

But when users have grossly incorrect mental models they are unlikely to be able to use the functionality effectively, as we saw in our Dropbox example. They may suffer serious losses, or be so nervous of making a costly mistake that they use a tiny subset of functions.

A bad solution to this problem is to try and educate users. It won't work in general because most users will resist spending time learning how to use an app; they expect (not unreasonably) to pick it up as they go along. A better solution is to design the concepts of the software so that they are simple, flexible and well suited to the user's needs, and to design the user interface to convey those concepts to the user.

The concepts themselves are then the basis both for the intended mental model of the user and the specification given to the programmer. The task of the user interface designer is to project what the usability researcher Don Norman calls a "system image" that corresponds faithfully to the conceptual model, so that users will acquire a mental model aligned with it.[25]

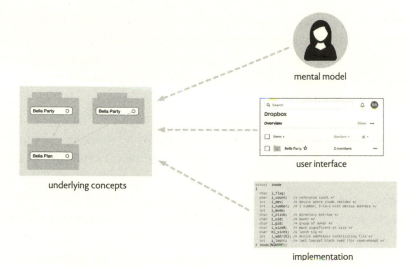

mental model

user interface

underlying concepts

implementation

FIG. 2.8 *The central role of concepts (left) in aligning the user's mental model (top right) and the developer's design model embodied in the code (bottom right). By mapping the concepts carefully to the user interface (middle right), the concepts are not only fully supported but also conveyed implicitly to the user.*

Figure 2.8 depicts this. At the top, there is the user; at the bottom, the code written by the programmer; and between the two, the user interface. For the software to be successful, we need to understand the user (by investigating her needs, working environment and psychological qualities); ensure that the code meets its specification (by testing, review and verification); and craft a usable interface. But most important of all is aligning the model in the head of the user with the model in the head of the programmer, and that is achieved by explicitly designing concepts that are shared by user and programmer alike, and conveyed clearly in the user interface.

Lessons & Practices

Some lessons from this chapter:

· Major usability problems in software applications can often be traced to their underlying concepts. In Dropbox, for example, confusions over whether deletions will affect other users is explained by Dropbox's adoption of a concept that originated in Unix.

· Software design is conducted at three levels: the physical level, which involves designing buttons, layouts, gestures and so on that are matched to the physical and cognitive capabilities of human users; the linguistic level, which involves designing icons, messages and terminology to communicate with users; and the conceptual level, which involves designing the underlying behavior as a collection of concepts. The two lower levels are concerned with representing the concepts in the user interface.

· For users, having the right mental model is essential for usability. To ensure this, we need to design concepts that are simple and straightforward, and to map the concepts to the user interface so that the concepts are intelligible and easy to use.

And some practices you can apply now:

· Take an app you have trouble using. Ask yourself what concepts are involved, and check that your hypotheses about how they work match the actual behavior. If not, can you find different concepts that explain the behavior more accurately?

· As the designer of an app, consider the functions that users find hardest to use (or easiest to misuse). Can put your finger on one or more concepts that are responsible?

· When designing, know what level you're working at. Start at the conceptual level, and move down. Sketching concepts out at the lower levels can help you grasp them more intuitively, but resist the temptation to polish the physical interface (for example, worrying about typefaces, colors and layout details) before you have a clear sense of your concepts.

· When you hear complaints about an app that focus on physical or linguistic aspects, ask if the underlying issue may lie at the conceptual level instead.

3
How Concepts Help

In traditional design disciplines, design evolves from a conceptual core. This core differs from field to field. Architects call it the *parti pris*: an organizing principle for the work that follows, represented by a diagram, a short statement or an impressionistic sketch. Graphic designers call it *identity*, and it typically comprises a few elements that capture the spirit of the project or organization. Composers build music around *motifs*—sequences of notes—that can be altered, repeated, layered, and sequenced together to form larger structures. Book designers start from a *layout* that specifies the dimensions of the text block and margins, and the typefaces and sizes in which the text will be set.

When the core is well chosen, the subsequent design decisions can seem almost inevitable. The design as a whole emerges with a coherence that makes it look like the product of a single mind even if it was the work of large team. Users perceive a sense of integrity and uniformity, and the underlying complexity gives way to an impression of simplicity.

For a software application, the conceptual core consists of—no surprise here—a collection of key concepts. In this chapter, we'll explore the roles that such concepts play, such as characterizing individual applications, application families, and even entire businesses; exposing complexity and usability snags; ensuring safety and security; and enabling division of labor and reuse.

Concepts Characterize Apps

If you're trying to explain an app, outlining the key concepts goes a long way. Imagine encountering someone who time-traveled from the 1960s and wanted to know what Facebook (Figure 3.1) was and how to use it. You might start with the concept of *post*, explaining that people author short pieces that can be read by others; that these are called "status updates" in Facebook (and "tweets" in Twitter) is a small detail. Then there's the concept of *comment*, in which one person can write something in response; the concept of *like* in which people

FIG. 3.1 *A screenshot of Facebook in which three concepts are evident: post (represented by the message and the associated image), like (represented by the emoticons at the bottom left), and comment (represented by the link on the bottom right).*

can register approval of a post, purportedly to boost its display ranking; and of course the concept of *friend* that is used both to filter what's shown to you and to provide access control so you can limit who sees your posts.

The difference between apps that offer similar functionality can often be explained by comparing their concepts. For example, a key difference between text messaging and email is that text messages are organized using a *conversation* concept in which all messages sent to a particular recipient appear; email messages, in contrast, are typically organized using concepts such as *mailbox, folder* or *label*. This is partly because the senders and recipients of text messages are uniquely identified by their phone numbers, whereas email users tend to have multiple addresses, which makes grouping into conversations unreliable. It also reflects different modes of interaction, with text messages relying on the context of the conversation and email messages more often interpreted in isolation (and thus often quoting previous messages explicitly).

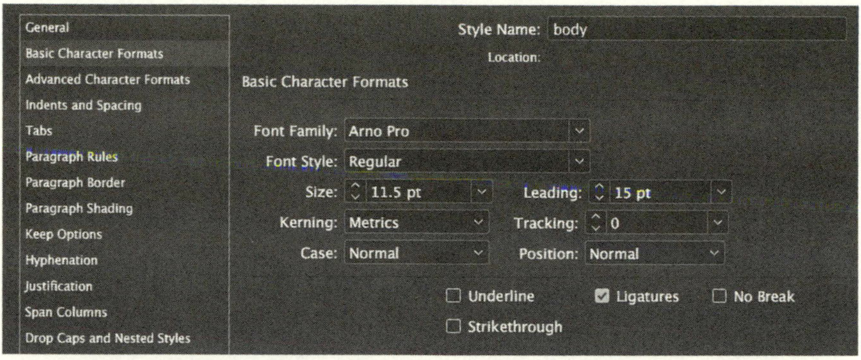

FIG. 3.2 *The style concept, in Adobe InDesign, showing one tab of the formatting settings for a style called "body," which is the style associated with regular paragraphs in this book.*

Sometimes it takes experience and expertise to identify the key concepts in an app. Novice users of Microsoft Word, for example, might be surprised to learn that its central concept is *paragraph*. Every document is structured as a sequence of paragraphs, and all line-based formatting properties (such as leading and justification) are associated with paragraphs rather than lines. If you want to write a book in Word, you won't find any concepts that correspond to its hierarchical structure—no *chapter* or *section*, for example—and headings are treated as paragraphs like any other. Word achieves its flexibility and power through the *paragraph* concept and by the powerful way in which it is combined with other concepts.[26]

Concepts Characterize Families

Concepts not only distinguish individual apps, but also unify families of apps. Programmers, for example, commonly use *text editors* (such as Atom, Sublime, BBEdit and Emacs) to edit program code; people use *word processors* (such as Word, OpenOffice and WordPerfect) to create documents of all sorts; and professional designers use *desktop publishing* apps (such as Adobe InDesign, QuarkXPress, Scribus and Microsoft Publisher) to organize documents into finalized layouts in books and magazines.

The key concepts of text editors are *line* and *character*. The *line* concept embodies both powerful functionality (such as the ability to perform "diffs" and "merges" which are essential to programmers for managing code) as well as limitations (notably that there is no distinction between a line break and a

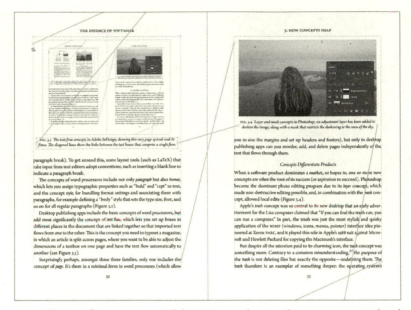

FIG. 3.3 *The text flow concept, in Adobe InDesign, showing this very page spread and its flows. The diagonal lines show the links between the text boxes that comprise a single flow.*

paragraph break). To get around this, some layout tools (such as LaTeX) that take input from text editors adopt conventions, such as inserting a blank line to indicate a paragraph break.

The concepts of word processors include not only *paragraph* but also *format*, which lets you assign typographic properties such as "bold" and "12pt" to text, and the concept *style*, for bundling format settings and associating them with paragraphs, for example defining a "body" style that sets the type size, font, and so on for all regular paragraphs (Figure 3.2).

Desktop publishing apps include the basic concepts of word processors, but add most significantly the concept of *text flow*, which lets you set up boxes in different places in the document that are linked together so that imported text flows from one to the other. This is the concept you need to typeset a magazine, in which an article is split across pages, where you want to be able to adjust the dimensions of a textbox on one page and have the text flow automatically to another (see Figure 3.3).

Surprisingly perhaps, amongst these three families, only one includes the concept of *page*. It's there in a minimal form in word processors (which allow

FIG. 3.4 *Layer and mask concepts in Photoshop; an adjustment layer has been added to darken the image, along with a mask that restricts the darkening to the area of the sky.*

you to size the margins and set up headers and footers), but only in desktop publishing apps can you reorder, add, and delete pages independently of the text that flows through them.

Concepts Differentiate Products

When a software product dominates a market, or hopes to, one or more new concepts are often the root of its success (or aspiration to succeed). Photoshop became the dominant photo editing program due to its *layer* concept, which made non-destructive editing possible, and, in combination with the *mask* concept, allowed local edits (Figure 3.4).

Apple's *trash* concept was so central to its new desktop that an early advertisement for the Lisa computer claimed that "If you can find the trash can, you can run a computer." In part, the trash was just the most stylish and quirky application of the WIMP (windows, icons, menus, pointer) interface idea pioneered at Xerox PARC, and it played this role in Apple's 1988 suit against Microsoft and Hewlett Packard for copying the Macintosh's interface.

But despite all the attention paid to its charming icon, the *trash* concept was something more. Contrary to a common misunderstanding,[27] the purpose of the *trash* is not deleting files but exactly the opposite—undeleting them. The *trash* therefore is an exemplar of something deeper: the operating system's greater tolerance to user errors, now recognized as a fundamental principle of user interface design. (More on the *trash* concept in Chapter 4.)

The spreadsheet, invented by Dan Bricklin in 1979 and one of the most successful innovations in computing, brought a new model of computation inspired by accounting ledgers. But its crucial novelty was not an accounting feature, but a remarkable new concept: the *formula*, which allowed the value of one cell to be defined in terms of values in other cells. In fact, VisiCalc, Bricklin's product, was not an accounting app at all, and other apps that had been targeted directly at accounting failed. The *formula* concept was powerful because it allowed you to model all kinds of computations. It was also nontrivial, relying on the subtle partner concept of *reference* which, by distinguishing absolute and relative positions, made it possible to copy a formula from one cell to another.

As a more contemporary example, the scheduling company Calendly offers an app whose differentiator is a concept called *event type*. In short, you define a collection of types of events (such as 15 minute phone calls, hour-long in-person meetings, etc), each with its own characteristics, such as the length of the event, its cancellation policy, what kinds of notifications occur, and so on. Then you indicate what your availability is for each event type, and people book appointments based on the event types.

It's a fun and enlightening game to try and identify the linchpin concept in a familiar app or system. Take the World Wide Web, for example. You might guess *html* or *link*, but markup languages and hypertext had been around for a long time. The core of the Web is in fact the *url* (uniform resource locator) concept: the idea of giving globally unique and persistent names to documents. Without that, the Web would be no more than a collection of proprietary networks, each operating in its own silo.

Concepts Expose Complexity

Many concepts are straightforward and easily learned. But others are more complicated. Some complications are not warranted and are just evidence of bad design (more on this below), but sometimes the power that a concept brings justifies its complexity.

Photoshop's *layer* and *mask* concepts fall in this category. When I started with Photoshop, I tried to learn it by playing around and by watching videos showing you how to do specialized tasks such as removing red-eye. Eventually, though, I realized that I needed to understand the core concepts more deeply, so I found a book that explained layers and masks (and channels, curves, color spaces, histograms, etc.) from a conceptual point of view, and I was then able to do whatever I wanted.

Some of the most complex concepts appear in apps that are widely used by non-experts. Browser apps include the *certificate* concept for checking that the server you're talking to belongs to the company you expect—your bank, for example, rather than an interloper trying to steal your credentials—and offer the *private browsing* concept to prevent your browsing information from being available to others after you've logged out. Despite the critical importance of these concepts for security, they are poorly understood. Most users have no idea how certificates work and what they're for, and they often think that private browsing allows them to visit sites without being tracked.

Worse, some of the most basic behaviors of browsers rely on complex concepts that are invisible to most users. The *page cache* concept, for example, is used by website developers to make pages load more quickly, by using previously downloaded content. But the rules for when old content is replaced (and how these rules are modified) are obscure even to some developers, so users and developers alike may be uncertain about whether content that appears in the browser is fresh or not.

Highlighting tricky concepts is helpful because of the focus it brings. It tells us, as users, what we need to learn: if you want to be a power user, just ignore all the details of the interface—those will come easily later—and master the handful of key concepts. It helps us, as teachers, to focus on the essence: so when we teach web development, for example, we can explain the important concepts—sessions, certificates, caching, asynchronous services, etc.—without getting caught up in the idiosyncrasies of particular frameworks. And it suggests opportunities to us as designers for innovation. A better concept for server authentication, for example, might prevent a lot of phishing attacks.

Concepts Define Businesses

"Digital transformation" is a grandiose term for a simple idea: taking the core of a business and putting it online, so customers can access services through their devices. In my experience as a consultant, I've sometimes found that executives, seeking to refresh and expand their business, instead of trying to understand what that core is, look to cool technologies. They hope to gain market share by moving to the cloud, or by incorporating machine learning or block-chains, often without a clear sense of the problem being solved.

Investing in core concepts is less flashy but likely to be more effective. First, just identifying the core concepts of a business helps you focus on what services you provide now (and might provide in the future). Second, analyzing those concepts can expose friction and opportunities to streamline the business. Third, an inventory of concepts can be ranked to reflect the value of each concept (to the customer and the company) and the cost of implementing and maintaining it, providing a basis for strategizing about the services the company provides. Fourth, by consolidating a set of core concepts, a company can ensure that customers have a uniform experience across technology platforms and company divisions, and can reduce the cost of having multiple variants of concept, each with its own implementation (and the concomitant headache of resolving differences in schemas when data is transferred between parts of the company).

The best services revolve around a small number of concepts that are well designed and easy for customers to understand and use, and their innovations often involve simple but compelling new concepts. In Apple's concept of *song*, for example, Steve Jobs saw the opportunity to provide every step of the experience of selecting, buying, downloading and playing music in a single unified concept.[28]

In contrast, consider what the airline business offers. Its key concept is arguably the *seat*, and yet few concepts are as obscure and hard to use. In an attempt to maximize profit, most airlines hide their pricing strategies for seats (so that only experts can tell how the price they are being offered compares to the price of other seats on the same plane, being sold now or in the past), divulge little information about the details of the product (such as how much room you get, and how different seats on the same plane compare), and may not even let you choose your seat in advance. The *frequent flyer* concept is typically mired in caveats and exclusions, and often employs a collection of misleading and dishonest tactics to ensure that customers derive as little value as possible.[29]

FIG. 3.5 *Labels in Gmail. I've entered a search query to show messages with no user-defined labels, but the first item seems to carry such labels ("hacking" and "meetups"). The explanation is that Gmail shows all conversations that contain messages satisfying the query, along with any labels attached to their messages. So a conversation containing one message without labels, and another message with labels, will appear in this search.*

Concepts Determine Costs and Benefits

When you're planning the development of an app, you can use the list of candidate concepts to scope the app's functionality, and make trade-offs between costs and benefits. Of course, developers have been doing something like this for decades, using informal notions of functionality or features. What concepts bring is a clearer division of functionality into free standing units, each with its own value and costs.

Put another way, before including any concept in your design, you want to *justify* its inclusion by considering (a) the purpose of the concept (and how valuable that is to your users); (b) the concept's complexity (and thus the cost of developing it, and the cost to users in terms of potential confusion); and (c) the concept's novelty (and thus the risk it entails).

No doubt the 80:20 rule applies to concepts: that 20% of concepts deliver 80% of the benefit. That does not mean that the less useful concepts are not important. Often, a concept that is useless to one user is essential to another. But sometimes a concept that is central to the design of an app ends up being underused.

The *label* concept in Gmail, for example, is the key mechanism for organizing messages. It must account for a considerable portion of Gmail's development complexity. As we'll see later in the book, the concept is mired in complications, and seems to be a source of confusion to users (Figure 3.5). It's also the reason that Gmail users can't always distinguish which messages they've sent and which they've received, because the *sent* label, like all others, is attached in the display to an entire conversation and not to an individual message. And yet it seems that less than a third of Gmail users create any labels at all![30]

Concepts Separate Concerns

What is the single most important strategy in problem solving? The prize, I believe, must go to *separation of concerns*, in which different aspects or "concerns" are addressed separately, even if they are not entirely independent.[31]

Concepts provide a new way to separate concerns in software design. Suppose, for example, you're designing a group forum in which members post messages and share various kinds of assets (such as images). At first sight, you might identify a *group* concept that embodies all the behavior of a group, such as joining the group, posting messages, reading other people's posts, and so on. But a more granular design might separate the functionality into several smaller concepts: a simpler *group* concept to capture the membership aspect and the way in which assets (such as messages and posts) are associated with groups; a *post* concept for the composing and formatting of messages; an *invitation* concept for inviting members to join; a *request* concept so that members can initiate membership; a *notification* concept that governs the messages members receive to notify them when certain status changes occur or when other members respond to their posts; a *moderation* concept for handling moderation of messages; and so on.

Such a separation is potent because it allows the designer to focus on one aspect at a time: you don't need to be thinking about whether membership invitations can be revoked at the same time that you're trying to shape the moderation pipeline. Each concept can be enriched arbitrarily, even becoming a small system in its own right—or it might be omitted entirely if the designer decides that its cost is not in proportion to its benefit.

The same breakdown that benefits a single designer applies to a team. By assigning concepts to different team members (or subteams), you can make progress in parallel. With a distinct purpose for each concept, the individual efforts are unlikely to conflict, and any incompatibilities between the designed concepts can be resolved when they are composed together.

Concepts Bring Reuse

Breaking a design down into the most basic, elemental concepts exposes opportunities for reuse. In the group forum app, the designer, having identified the *moderation* concept, for example, would be wise to explore how moderation is achieved in other contexts. That posts in a forum are the subject of the moderation, rather than comments on a newspaper article, can initially be ignored. Once the variety of standard moderation options are ready to be considered, the designer can ask whether one might be more suited to the context than another. Even better, the designer might discover that an off-the-shelf solution can be adopted in its entirety, thus reusing not only the idea of the concept but its implementation too.[32]

Many concepts are reused in almost an identical form across apps. Imagine a handbook of concept designs. Rather than reinventing the wheel, a designer could look up a concept, and read about all the tricky issues associated with it, along with their conventional solutions.

For example, almost all social media apps incorporate some form of the *upvote* concept, in which users register their like (or sometimes dislike) of an item, with an impact on its prominence in feeds and search results. If you were designing this concept for the first time, you would probably realize early on that you needed some way to prevent double voting, and that this would depend on somehow identifying the user when they upvote an item.

But you might not know the range of ways to identify users and their relative advantages and disadvantages: whether you should make them log in and rely on their user names; or use their IP addresses as proxies; or install a cookie just for this purpose; and so on. You might not have thought to mitigate the cost of storing the identity of the user with each vote by freezing voting on archived items (so that the identities associated with prior votes can be discarded). Nor might you have considered weighting of votes: whether some users' votes should be more influential, or whether more recent votes should count for more than older votes. A handbook of concepts would have an entry for *upvote* that lists all these design options and their trade-offs, allowing you to save the effort of exploring design paths that others have traveled down many times before.

Concepts Help Identify Usability Snags

Sometimes a software app or service turns out to be very hard to use, frustrating some users so much that they reject it in its entirety. When this happens, the problem can sometimes be pinned on a single concept.

Apple offers cloud storage for data held on a user's devices: laptops, phones, etc. The cloud storage serves two distinct purposes. One is to synchronize data between devices. This makes it easy, for example, to maintain the same set of bookmarks in your browser irrespective of the device you're using. The other is to provide backup, so that if a device is lost or its storage damaged, the data can be restored from the copy held in the cloud.

Apple's design strategy has always favored simplicity and automation over manual control, even in cases where giving more control to the user seems essential. Its *synchronization* concept is a prime example of this strategy, and as a result is often the source of complaints from confused customers.

Sometimes Apple's design seems to put the user in a catch-22 situation without any viable options. Consider the dilemma you face if your iPhone runs out of storage space. When this happens, you get a warning message telling you that your phone is about to run out of space, suggesting that you "manage your storage in Settings."

At this point, none of your options are very attractive. Suppose you find that your photos are taking much of the space. You can delete the entire Photos application, with all its associated data and state. You can turn on "optimized storage" which converts photos to lower-quality versions (while keeping the higher-quality versions as master copies in the cloud). What you can't do, however, is simply to delete some of your photos from the phone, in the hope that they will remain on your other devices. If you do that, the photos you delete on your phone will also be deleted from the cloud. And then those deletions will be propagated to all your other devices, removing the photos from them too.

What's missing from Apple's *synchronization* concept is the idea of *selective synchronization*: the ability to indicate that some files should not be synchronized. With this feature, you would be able to delete old photos from your phone, while leaving copies in the cloud. Dropbox's *synchronization* concept, in contrast, does offer this feature, so it's certainly feasible.[33]

Concepts Ensure Safety and Security

The popularity of the term "secure by design" reflects a growing consensus that security—a concern for all software systems today—is best achieved not by closing all security holes (an impossible task anyway) but by designing systems to be secure despite them.

Security design relies on a few key concepts to establish system-wide properties: *authentication*, to ensure that the actors (or "principals" as they are called in the security field) who make requests are correctly identified; *authorization*, to ensure that those actors can only access certain resources; *auditing*, to ensure that every access is faithfully recorded (and bad actors can be punished accordingly); and so on.

Each of these concepts has many variants, and understanding the security of a system rests on understanding these variants deeply. If a concept is used casually, without carefully analyzing its purposes and assumptions, a system that appears to have all the right protections may turn out to be vulnerable.

Take *two-factor authentication*, for example. It works like this: a user logs in to a service, which then sends a special key to the user through another channel, usually by a text message to a phone; the user enters the key, and in return receives a credential (in the form of a cookie or some other kind of token) that grants the desired access. The user is confirmed by this scenario as the owner of the phone (say) and thus also as the legitimate account holder.

This design, however, is fraught with complexities. First, the assumption that access to a phone number implies ownership of the associated phone is questionable: Jack Dorsey, CEO of Twitter, fell victim to a "SIM swapping" attack in 2019 when hackers took control of his phone number. Second, the design involves an additional concept—namely *capability*, in which a token is provided that grants access to anyone who holds it—and in the interaction between the two concepts lies a gaping security hole.

Suppose you receive a phishing email, asking you to confirm a LinkedIn connection, containing a URL that points not to the real LinkedIn site but to a server owned by a hacker. This server mimics the real server, and gives you the impression that you are talking to LinkedIn itself. When you enter the two-factor authentication key, the malicious server passes it on to LinkedIn, obtains your access token, and passes the token on to you. All seems fine, but the hacker has the token too, and can now access your account as you.[34]

These issues are contained within a few critical security concepts. Only the design of the concepts and their interactions is relevant; the code must implement the concepts correctly, of course, but the design problems are fundamental. The security of a system will thus often rest on understanding its security concepts and their known vulnerabilities, and if an analysis suggests that stronger guarantees are needed, the concepts will need to be replaced or augmented. In short, security design is in large part about the design and use of appropriate concepts.

Concepts are central to the design of all critical systems.[35] In the domain of safety rather than security, there are fewer standard concepts. Nevertheless, recurring incidents suggest opportunities for new concepts that would play a role similar to concepts in security, embodying conventional ways to encapsulate critical functions. For example, medical devices are rife with mistakes related to dosage calculations, and a *dose* concept that handles all the complications associated with units, concentrations and flow rates might eliminate many of the tragic accidents in which patients are injured or killed by preventable errors.[36]

Concepts Ground Design Critique

In any field of design, critique—in which designers review and analyze each others' work—plays a central role. A critique (or "crit") is not a formal evaluation in which principles are systematically applied, and its very informality can provide space for insights and inspiration. Moreover, critique is inevitably subjective, as different participants come with their own biases and interests. But effective critique is always deeply informed by experience and expertise, which the critic brings by speaking in a language of known principles and patterns.[38]

Such principles and patterns have developed around the physical and linguistic levels of design, but not so much at the conceptual level. Even the widely accepted idea that a system should have a clear conceptual model is often interpreted as a principle at the linguistic level, focusing not on the structure of the conceptual model itself but on how faithfully and effectively the model is projected by the user interface.[37]

The rest of this book is aimed at filling exactly this gap. Part II provides a language for talking about concepts, and a structure for expressing them. Part III presents three design principles that govern the selection of concepts and their composition.

Design principles can be used in different ways. They offer a basis for shared understanding in design crits, or can be applied systematically in heuristic evaluations. But their more important role is in shaping the way designers think. Once you have grasped Norman's notion of "mapping," for example, which suggests arranging user interface controls in a layout that matches the layout of the objects being controlled, you'll intuitively create layouts that already have a natural mapping.[39]

Likewise, absorbing the language and principles of concepts should empower you to be a more effective software designer: it should give you a way to express your ideas more directly and clearly; to solidify your intuition and experience in a more systematic framework; and to inform your judgment with heightened design sensibilities.

Lessons & Practices

Some lessons from this chapter:

- Concepts characterize individual apps, app classes, and entire product families. They allow you to compare apps, highlight their essential capabilities, and learn how to use them effectively.
- Concepts are often product differentiators, can bring focus to marketing efforts, and can reveal the reason for a product's success or failure.
- Concepts can help a company engaged in "digital transformation" map a path forward. More than extending the platforms through which customers can access services or adopting hot technologies, digital transformation is about identifying, consolidating, and extending core business concepts, so that customers have a rich and uniform experience that brings real value.
- Concepts give a new granularity that lets software designers separate concerns, exploit reuse, and divide engineering labor more effectively.
- Concepts are the essence of design for safety and security, where picking the right concepts and understanding their implications is paramount.
- Concept design offers rules that can be applied when reviewing a design, to avoid problems that might otherwise not be discovered until much later; a designer who internalizes these rules is likely to produce better designs even if not considering them explicitly.

And some practices you can apply now:

- Take an app that you are familiar with and identify a small number of core concepts that characterize it. Consider similar apps, and see if you can explain the commonalities and differences in terms of their concepts.
- Try to identify the concepts that are responsible for the success (or failure) of software products that you have been involved in creating or using.
- Using the *group* concept as an example, find a complex feature in some software you're familiar with, and break it down into separate, smaller concepts. Does this reveal connections to other products, or opportunities to apply a concept more uniformly?

PART II
ESSENTIALS

4

Concept Structure

So far, I've talked in rather vague and general terms about concepts. But what exactly *is* a concept? To use concepts effectively, we'll need to go beyond generalities and look at specifics. In this chapter, I'll show you how to structure the definition of a concept. This structure will clarify what concepts are (and are not); it will give you a roadmap for designing concepts; and it will allow you to define each concept precisely.

I'll use three concepts as exemplars, concentrating on their structure but also telling the larger story of their invention and use, and pointing out some of their design subtleties.

Concepts don't make all design problems go away, of course. They do, however, help you *localize* challenges by recognizing them as specific to a particular concept. A concept becomes a container not only for the behaviors that it embodies but also for all the accumulated knowledge about its design, the issues that have arisen in practice as it has been deployed, and the various ways in which designers have dealt with them.

Apple's Killer Concept: Trash

The *trash* concept was invented by Apple for the Lisa computer, the predecessor of the Macintosh, in 1982. The trash icon (and its cute bulging version when you put something in it, and the nifty crunching noise it made when you emptied it) became the emblem of the Macintosh's claim to being a more friendly and usable operating system (Figure 4.1). Since then the *trash* concept has become ubiquitous, appearing not only in the file system managers of other operating systems, but also in many other applications.

At first glance, the trash icon appears to offer no more than an intuitive gesture for deleting files and folders—by dragging them to the trash rather than executing a more conventional deletion command. The real innovation, however, is not that you can move things *into* the trash, but that you can then move

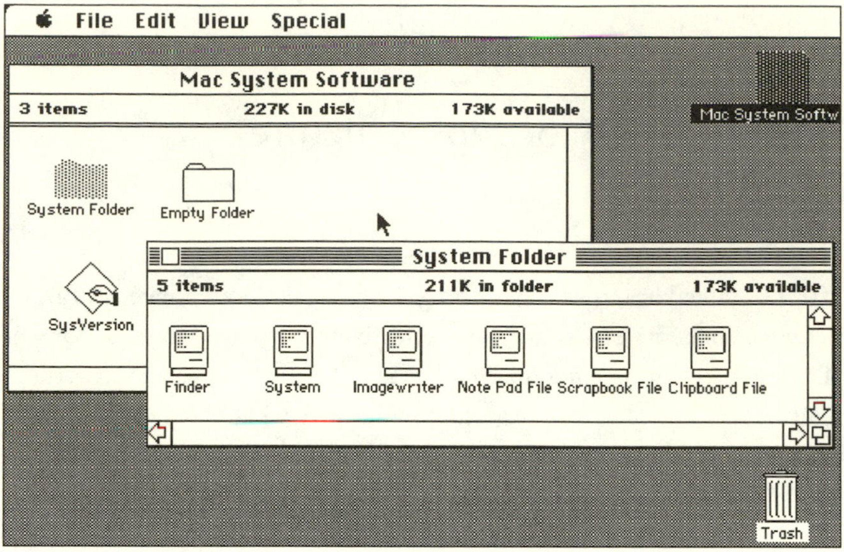

FIG. 4.1 *The original Macintosh desktop, with trash at bottom right (1984).*

them *out*. The trash holds a collection of items, which can be viewed by opening it; you can then restore items by moving them from the trash to some other location. We might therefore say that the *purpose* of the *trash* concept is not *deletion* but the *undoing* of deletion.[40]

Of course, you have to be able to delete files permanently too, in order to make space for new files. This is accomplished by "emptying" the trash. So, in summary, you move an item to the trash when you want to delete it; you take it out of the trash when you want to restore it; and you empty the trash when you run out of space and want to permanently remove the items it contains.

To make concept design practical, we need a way to describe concepts succinctly and precisely. Figure 4.2 shows how the *trash* concept can be described. The parts correspond exactly to the explanation I just gave you.

First comes the *name* of the concept.[41] Along with the name is a list of the types—here, just one type, *Item*—that can be specialized when the concept is instantiated. So in one instantiation, the items might become the files of a file system; in another, they might be the messages of an email client.

The name is followed by a pithy summary of the concept's *purpose*. Then comes the *state*, which organizes the things that comprise the concept into various structures. In this case, there are just two sets: *accessible*, the set of items that

```
1   concept trash [Item]

2   purpose
3      to allow undoing of deletions

4   state
5      accessible, trashed: set Item

6   actions
7      create (x: Item)
8         when x not in accessible or trashed
9         add x to accessible

10     delete (x: Item)
11        when x in accessible but not trashed
12        move x from accessible to trashed

13     restore (x: Item)
14        when x in trashed
15        move x from trashed to accessible

16     empty ()
17        when some item in trashed
18        remove every item from trashed

19  operational principle
20     after delete(x), can restore(x) and then x in accessible
21     after delete(x), can empty() and then x not in accessible or trashed
```

FIG. 4.2 *The trash concept defined.*

are accessible (outside the trash), and *trashed*, the set of trashed items, namely those that have been deleted but not yet permanently removed.[42]

The *actions* describe the dynamics, the behavior of the concept. Actions are instantaneous—that is, take no time—but any amount of time can pass between them. The description of an action says how the state is updated when the action occurs—for example, that deleting an item moves it from the set of accessible items to the set of trashed items. It may also include a precondition that limits *when* the action may occur—for example, that you can only delete an item if it is accessible but not trashed. (In addition to the actions that came up in the informal overview, the full description includes a *create* action for completeness, since the items that are deleted must somehow come into being.)

Finally, there's the *operational principle*, which shows how the purpose is fulfilled by the actions, and comprises one or more usage archetypal scenarios. In this case, there are two scenarios. One is for restoring: it says that after you delete an item x, you can restore it and that item is then accessible. The other is

49

for permanent removal: it says that after you delete an item you can empty the trash and the item will be neither accessible nor in the trash.

In a narrow technical sense, the operational principle doesn't add anything, because you can infer any scenario from the action specifications. But for understanding *why* the concept is designed the way it is, and *how* it's expected to be used, the operational principle is fundamental.[43]

This can all be made more precise, with a mathematical model of behavior, and a formal notation for defining actions and operational principles. The details won't matter to most readers, so I've relegated them to the endnotes.[44]

The Trash Concept: Design Flaws Finally Fixed

This *trash* concept has been very successful, and is widely used. It appears in all graphical file managers (Mac, Windows, and Linux), in email clients (such as Apple Mail and Gmail), and in cloud storage systems (such as Dropbox and Google Drive). Not all instances of the concept offer the exact same behavior; in one common variant, items are permanently removed from the trash when a certain amount of time (say, 30 days) has elapsed since they were deleted.

On the Macintosh, there is only one trash for the entire system, which has some unfortunate consequences. First, as you insert and remove external drives the contents of the trash grow and shrink as the deleted items belonging to those drives come and go. This can be mildly disconcerting: you might see an item in the trash and intend to restore it, but then find that it has disappeared (because the drive has been ejected).

A more substantive problem arises in the following scenario. Suppose you insert a USB key (a small external flash drive) and try to copy a file to it, but there's not enough room. So you decide to make some space by trashing some files already on the key. You attempt to copy the file to the key a second time, but it fails again. Trashing the files, you then realize, didn't actually make space by removing them permanently; to do that, you need to empty the trash.

But now you have a dilemma. If you don't empty the trash, you won't be able to copy the file to the USB key. But if you do, you'll lose all the files you previously deleted from your hard drive, eliminating the option to restore them in the future.

Amazingly, this problem was left unresolved for over 30 years, and was finally addressed in OS X El Capitan, the version of Apple's operating system issued

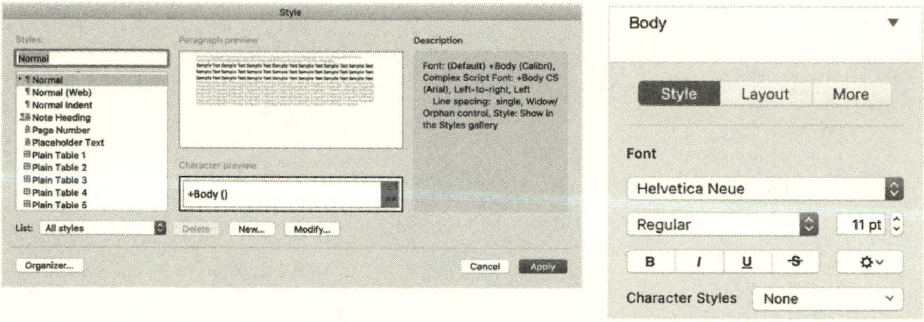

FIG. 4.3 *Style concept in Microsoft Word (left) and Apple Pages (right).*

in 2015. The resolution, which is perhaps more a workaround, was to offer a "delete immediately" action that allows you to permanently remove select items in the trash in one click.

Another design flaw involved how files in the trash were listed. For decades, there was no way to sort items in the trash by date of deletion. This meant that if you deleted a file accidentally, and then went to the trash hoping to restore it, you might be out of luck. If you wisely let the trash grow until you really needed the space, your trash would contain thousands of files. If you deleted a file and couldn't remember its name, you had no way to find it.

In 2011 (with OS X Lion), Apple let you sort items in a folder by their "added date," which, for the trash, corresponds to the date of deletion. (In Chapter 6, I explain this design in more detail, and show how it involves a clever fusion of concepts.)

The Concept Behind Desktop Publishing: Style

Our second example is the concept of *style*, previously mentioned and shown for Adobe InDesign in Figure 3.2 and for Microsoft Word and Apple Pages in Figure 4.3. Its purpose is to make it easier to achieve consistent formatting.

To use it, you assign styles to paragraphs in your document. For example, you might assign the style *heading* to each of the paragraphs that corresponds to a section heading. Now if you want to make all the headings bold, say, you change the format setting of the *heading* style to *bold*, and all the heading paragraphs will be updated in concert.

That is the operational principle. It's actually a rather elaborate scenario, which involved creating multiple paragraphs, assigning a single style to more

```
1   concept style [Element, Format]

2   purpose
3      easing consistent formatting of elements

4   state
5      assigned: Element -> one Style
6      defined: Style -> one Format
7      format: Element -> one Format = assigned.defined

8   actions
9      assign (e: Element, s: Style)
10        set s to be the style of e in assigned

11     define (s: Style, f: Format)
12        set s to have the format f in defined
13        create s if it doesn't yet exist

14  operational principle
15     after define(s, f), assign (e1, s), assign (e2, s) and define (s, f'), e1 and e2 have format f'
```

FIG. 4.4 *The style concept defined.*

than one of them, and then modifying that style. The operational principle is not always the simplest scenario, but it *is* the smallest one that demonstrates how the purpose is fulfilled. Obviously, to demonstrate consistent formatting of paragraphs, you need more than one paragraph. In the concept definition, the operational principle says that if you define a style *s* to have format *f*, assign it to two elements *e1* and *e2*, and then redefine *s* to have format *f'*, both *e1* and *e2* will now have that new format.

To make this magic work, the concept state (see Figure 4.4) has to be quite complicated. There are two mappings, one (which I've called *assigned*) that associates with each element the one style that is assigned to it, and the other (which I've called *defined*) that associates with each style the format that it defines. In this description, a "format" is an abstract thing which you can think of as embodying all the formatting properties (bold, 12pt, Times Roman, etc.). A third component of the state, called *format*, is introduced as a shorthand, and stands for the combination of these two mappings (written *assigned.defined*), so that if an element *e* is assigned a style *s*, and style *s* is defined to have format *f*, then element *e* has format *f*.[45]

I've defined two actions: one that assigns a style to an element, and one that defines the format for a given style. This second action is used both to create a style with a given format, and to update an existing style with a new format. These might have been listed as separate actions; both approaches are valid.

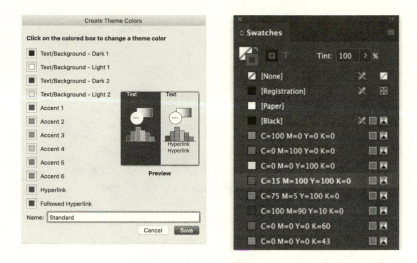

FIG. 4.5 *Style concept applied to slide themes in Microsoft PowerPoint (left) and color swatches in Adobe apps (right).*

Stylish Things: Not Always Genuine

The concept of *style* is widely used. It appears in word processors such as Microsoft Word and Apple Pages, and desktop publishing tools such as Adobe InDesign and QuarkXPress, not only for styling paragraphs but also for styling ranges of characters. It's used in Microsoft PowerPoint for "color themes," which comprise a collection of predefined styles for coloring the various kinds of text (e.g., titles, hyperlinks, body text) and backgrounds that appear on slides (Figure 4.5, left). The "classes" of the web's formatting language, Cascading Style Sheets (css), are styles too, where they provide a clean separation of formatting from content.

Sometimes the use of the style concept is not immediately apparent. In Adobe InDesign and Adobe Illustrator, you can color elements by applying a color swatch (Figure 4.5, right). What you might not notice at first is that the color swatches are modifiable. If you color several elements with a red swatch, they will all become red—no surprise there. But now if you open up the red swatch and adjust its sliders to make it green instead, you'll see that all those red elements now become green. This is tremendously useful of course, because it

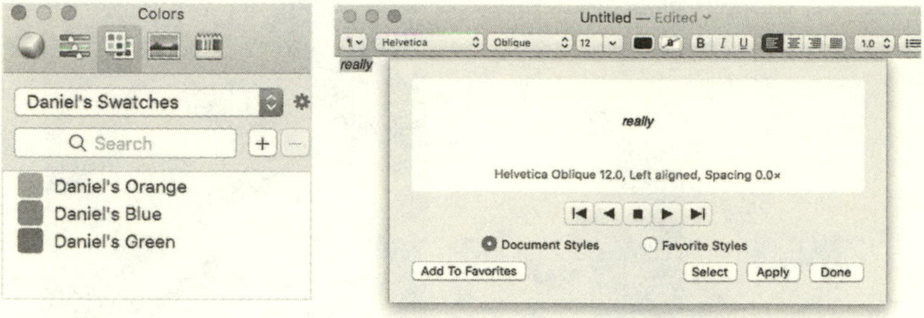

FIG. 4.6 *Two concepts that look like style but are not: the Apple color picker (left) and styles in Apple TextEdit (right).*

makes it easy to maintain a consistent palette, without having to settle on the palette's colors at the start.

Other times, you come across something that appears to be an instance of a concept but turns out not to be. Apple's color picker, used in all of Apple's apps for selecting colors, looks very similar to the Adobe color swatches (see Figure 4.6, left), so we might guess that it's an instance of the *style* concept. But if you play with it, you'll discover that while you can delete a swatch or add a new one, you cannot *change* the color of an existing swatch. This ability to modify the format of a style is essential to the *style* concept. Without it, the concept cannot work at all—that is, its operational principle fails. Adding a new style has no effect on elements associated with an old style, so there's no way to change elements in concert once their format has been defined.

Another near miss is the "style" of Apple's rudimentary word processor, TextEdit (Figure 4.6, right). The name suggests the *style* concept, and you can indeed not only create and delete named "styles" but you can modify them too. When you apply a style to a paragraph, however, it updates the formatting of the paragraph, but the style does not stick: there is no persistent association of the style with the paragraph. Changing a style thus only impacts its future uses, and not the paragraphs that it was applied to before.[46]

The *style* concept has been enriched in various ways, many of which involve layering of formats one on top of another: for example, partial styles, whose formats only set some properties (say making text italic but not affecting its size); style inheritance, in which one style is defined as an extension of another; and overrides, in which the format of an element is defined by a style with some overriding format.

```
1   concept reservation [User, Resource]

2   purpose
3     manage efficient use of resources

4   state
5     available: set Resource
6     reservations: User -> set Resource

7   actions
8     provide (r: Resource)
9       add r to available

10    retract (r: Resource)
11      when r in available and not in reservations
12      remove r from available

13    reserve (u: User, r: Resource)
14      when r in available
15      associate u with r in reservations and remove r from available

16    cancel (u: User, r: Resource)
17      when u has reservation for r
18      remove the association of u to r from reservation and add r to available

19    use (u: User, r: Resource)
20      when u has reservation for r
21      allow u to use r

22  operational principle
23    after reserve(u, r) and not cancel(u,r), can use(u, r)
```

FIG. 4.7 *The reservation concept defined.*

A 19th Century Concept: Reservation

For our last example in this chapter, let's consider a familiar concept that existed long before software. The *reservation* concept (Figure 4.7) helps to make efficient use of a pool of limited resources. The provider wants the resources to be used as much as possible; the consumer wants a resource to be available when they need it.

It works like this. A consumer wanting to use a resource attempts to make a reservation. If there isn't already a reservation for that resource, the attempt succeeds. Then later when the consumer comes to use the resource it will be available to them.

To run this concept, you need to track the set of reservations, each being associated with the resource being reserved and the consumer who reserved it. In addition to making a reservation and then eventually using the resource, it's usually also possible for the consumer to cancel the reservation if they decide they don't need it after all.

Of course, this is belaboring the obvious. You've reserved tables at a local restaurant, books at a library, seats at a concert, etc., so none of this is new. But it's worth noting the form the explanation took. I started with the purpose (making efficient use of resources); then I gave you the operational principle (how a reservation is made and fulfilled); then the state (a set of reservations); and then finally the actions (*reserve*, *use* and *cancel*).

The description of this concept uses a set (to remember which resources are still available) and a mapping from user to resources for the reservations. This mapping, unlike the mappings in the definition of the *style* concept, is one-to-many: a single user can reserve multiple resources.

The actions also include those performed by the owner of the resources (the restaurant, for example), for providing and retracting a resource. Retracting a resource is tricky if it's been reserved. For simplicity, the definition says that you can't retract in this case, but in practice a better design would allow the re-traction, with an implicit canceling of the reservation. Finally, the operational principle adds a caveat that was missing in my informal account: you can only use the reservation if, having made it, you don't then cancel it before you use it.

A Designer's Reservations

Like any other concept, *reservation* has many variants and additional features. Often the resource is tied to a period of time. In a restaurant reservation system, the consumer chooses only the start time and the end time is implicit, deter-mined by the restaurant owner (a tricky business, because setting dining times that are too long will mean serving fewer customers, but setting them too short will result in making people with reservations wait). The resource may be asso-ciated with a particular physical object (a seat on a plane, for example) or it may represent a fungible member of a class (such as a table anywhere in a restaurant, or any copy of a particular book).

Because reservations are often free, the provider may need to protect against users who reserve a resource but never use it. Restaurant reservation systems do this by adding a *no-show* action that is executed by the restaurant owner; if a customer has too many no shows, their account is suspended. Another strategy is to prevent consumers from reserving resources that cannot be used together, such as booking tables at two different restaurants on the same night. Airlines have complex rules for detecting conflicting reservations, which lead to strange anomalies.[47]

The *reservation* concept is so useful that it has applications in many different domains. In railway signaling, safety is achieved by requiring trains to reserve segments of track before entering them; that way, the system can ensure that no two trains have ever occupy the same segment at the same time. And in networking, there is a protocol called RSVP (Resource Reservation Protocol) that allows routers to reserve bandwidth so they can guarantee a certain level of network performance (called "quality of service") when they need it.

Lessons & Practices

Some lessons from this chapter:[48]

- A concept definition includes its name, purpose, state, actions and operational principle. The operational principle, showing how the behavior fulfills the purpose, is the key to understanding a concept, and may not be the simplest scenario.
- Every concept was invented by someone at some time, for some purpose. Most concepts in widespread use have undergone extensive development and refinement over time.
- Most concepts are generic, and can be applied in many different contexts on different kinds of data. Genericity affords reuse, and helps distill concepts to their essence.
- Concepts can be designed and understood independently of one another, simplifying software design by allowing a design to be broken into distinct subproblems, many of which will be solvable by reusing existing concepts.

And some practices you can apply now:

- To design or analyze a software product, start by identifying the concepts. For each concept, choose a good name, find a pithy summary of the purpose, and formulate an operational principle. To delve more deeply, list the actions, and figure out what state is needed to support them.
- If your concept lacks interesting behavior—you can't come up with a compelling operational principle, or can't even list actions—it may not be a concept at all, or you may need to expand it to identify the real concept.

- Take a database schema or class structure for a system, represent it as an entity-relation diagram, and then break it into smaller diagrams (overlapping on the entities but not the relations), each of which embodies some piece of functionality. These smaller diagrams are the states of distinct concepts.
- Conversely, to design a data model, rather than treating it as a monolith, develop local "micromodels" for each of your concepts independently, and then merge them on the common entities to form the global model.
- As a fun and rewarding exercise, pick an interesting concept and research its history. It might be a concept (like *reservation*) that exists independently of software, or it might be a concept in your favorite app. When was it invented and by whom? How has it changed over time?

5

Concept Purposes

Purposes matter in all aspects of life because they help us set direction, explain ourselves to others, and reach consensus in collaborations. Design is no different from any other activity in this respect: you can't design something well without knowing *why* you want it in the first place.[49]

For concepts, purposes are essential too. For a designer, the purpose is what justifies the effort of designing and implementing the concept. For a user, it tells you why you might want it—and if you don't know what something is *for*, it's hard to understand *how* you would use it.

Before you dismiss this as obvious, consider that the designers of software rarely articulate purposes beyond those of the product as a whole. I'm proposing a more radical idea here. It's not enough to know why you're designing a product. You need to have a *why* for each element of your design, a purpose for every concept.

Finding purposes for concepts is hard work, but it's amply rewarded in the insights it gives you about the problem you're trying to solve, and in forcing you to focus on what matters.[50] In this chapter, we'll see how subtle it can be to figure out the purpose of a concept, and how failing to formulate or convey a straightforward purpose can have nasty consequences. In software especially, with its capacity for boundless complexity, it's easy to get mired in details and lose track of the big picture. Thinking about purposes helps you draw back and regain your bearings.

Once you have a purpose, you can ask whether your concept fulfills it. As we'll see, that's not always straightforward, because a purpose isn't a simple description of an expected behavior. It's an articulation of a need, and that need can vary depending on the user and the context of use. *Misfits*, in which forms fail to fit their context—or in our case, concepts fail to fulfill their purposes—are often unpredictable, because neither needs nor contexts of use can be fully anticipated at design time, let alone reduced to precise logical statements.

Concepts don't eliminate this problem; their value is that they give you a framework that mitigates it, by elevating the role of purposes, and by providing a structure for organizing the knowledge accumulated through experience of design and use. In this chapter, I'll show some of the benefits of thinking in terms of purposes; later, in Chapter 9, I'll revisit the relationship between purposes and concepts and refine some of these ideas.

Purposes: First Step to Clarity

To be easy to use, a concept must have a clear purpose. And the purpose can't be the designer's secret; it must be shared with the user.

When I upgraded to the most recent version of Apple Mail, I noticed a new vip button, and looked it up. Here's what Apple's help guide said:

> *You can easily keep track of email messages from people important to you by making them VIPs. Any messages in the Inbox from a VIP (even sent as part of a conversation) are displayed in a VIP mailbox …*

In just two sentences, they've given me the purpose of the *vip* concept (tracking emails from people important to me), and most of the concept's operational principle (it seems I can make someone a VIP, and then, subsequently, messages from them will appear in a special mailbox).

In contrast, I wanted to understand what a section is in Google Docs, so I looked up the word "section" in the online help. The first disconcerting thing was that no article about the *section* concept appears. The closest was an article entitled "Work with links, bookmarks, section breaks, or page breaks." I scrolled through that article and found this:

> *If you want to break up ideas or set images apart from text in your document, you can add section or page breaks in Google Docs.*

That's it, as far as explaining the purpose goes. I might have guessed that sections could be used to break up ideas, so that doesn't help me. The reference to images made me a bit nervous, because it seemed to be suggesting that I couldn't have images that are "apart" from my text without having sections. In short, it left me no wiser about what sections are for.

It turns out that the purpose of *section* is to allow different parts of your document to have different margins, headers and footers, and to have subsequences of pages with their own page numberings. You *can* have images separate from text without using any sections; what sections let you do is change the margins of an image independently of the margins of the text around it.

Criteria for Purposes

Defining compelling purposes for concepts is hard. Because purposes are about human needs in a human context, they cannot be assessed in a logical or mathematical way, but must always be informal and rough-edged. Nevertheless, here are some criteria that can help:

Cogent. A purpose must be a cogent expression of an intelligible need, and not just a vague hint about some desires the user might have or tasks she may want to perform. The phrases used in the explanation of the purpose of the Google Docs *section* concept—"break up ideas" and "set images apart from text"—give us only a hazy idea of what the user is trying to do. In contrast, "allowing different margins on different pages" is pretty clear.

Need-focused. A purpose must express a need of the user, and not just recapitulate a behavior whose significance is unclear. Take the *bookmark* concept that browsers offer. It's not helpful to say that the purpose of a bookmark is to "mark a page" or to "save a favorite page": these just beg the question of why you would want to do such things. Instead, the purpose might be to "make it easier to revisit a page later," or perhaps to "share a page with another user." Don't worry if a purpose isn't precise initially; if your formulation immediately raises questions (may "revisiting later" be on a different device?), that's a good sign of progress!

Specific. A purpose must be sufficiently specific to be relevant to the design of the concept at hand. You could say that any concept's purpose is to "make the user happy" or "let the user work more effectively," and such expressions are indeed cogent (since we know exactly what they mean) and need-focused. But clearly such purposes would not be a helpful basis for a concept design because they are not specific enough to distinguish one concept from another.

Evaluable. A purpose should provide a yardstick to measure a concept against. You should be able to take the operational principle and easily assess whether it fulfills the purpose. For the *trash* concept, the purpose "allow undoing of deletion" is clearly supported by the scenario in which a file is deleted and then restored from the trash. In contrast, the purpose "prevent accidental

FIG. 5.1 *Call forwarding: a design puzzle (top) and two solutions (below). If A is forwarded to B, and B is forwarded to C, does a call to A get forwarded to B or to C?*

removal of files" would not be very helpful, because we'd need additional information about user behavior, notably whether users might not only delete files accidentally but also empty the trash accidentally.

Purposes Resolve Design Puzzles

Sometimes, a design you're working on presents multiple options that seem equally plausible, and there seems to be no rational basis for selecting one over another. In many cases, this dilemma is a result of not understanding the purpose deeply. Once the purpose is understood, it becomes clear which option is the right one.

Take *call forwarding*, a telephony concept that allows calls to one line to be automatically transferred to another. Suppose we have three users with phone lines *A*, *B* and *C* (Figure 5.1). Now imagine that our first user forwards her line to *B*, so that calls to *A* are now redirected to *B*. What if the second user now forwards to *C*, so that calls to *B* go to *C*? If a call comes in to *A*, should it be forwarded to *B* (based on the first user's forwarding request), or should it be forwarded to *C* (based on the requests of both users)?

The solution to this dilemma is to recognize two distinct purposes for call forwarding. One, which might be called *delegate*, is to allow a person to delegate their calls to somebody else. In this case, if the owner of *A* has delegated to *B* and *B* has delegated to *C*, then clearly a call to *A* should be forwarded two steps

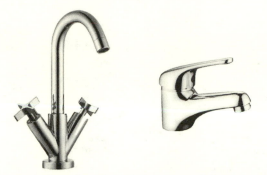

FIG. 5.2 *A physical analogy: do faucet controls have purposes?*

to C. The other, *follow me*, is to allow a person to transfer calls when they are working in a different location. In this case, if the owner of A has moved to the location of B, and the owner of B is now in the location of C, a call to A should clearly be forwarded only to B.[52]

Articulating these two distinct purposes reveals that there are two distinct concepts, *delegate forwarding* and *follow-me forwarding*, each serving its own purpose. Their behaviors may differ in other respects too. The *delegate forwarding* concept, for example, might support an option in which the call rings first at A and is only transferred to B if unanswered.

Concepts without Purposes: Faucets & Editor Buffers

A concept may have no compelling purpose at all. This casts some doubt on its utility, but people can find value even in dubious concepts. The *poke* concept was one of Facebook's earliest, and nobody ever really knew what it was for.[53]

Concepts more commonly lack purposes because their design was motivated not by a genuine user need, but because it was easier to build it that way. A physical analogy illustrates the point. Compare two kinds of common mixer faucets (Figure 5.2). In both kinds, there is a single outlet pipe that mixes the hot and cold water.

In the older kind (on the left), this pipe is fed by two separate faucets, marked "hot" and "cold." Viewed as concepts, these two faucets have no compelling purposes. It's clear what they *do*; opening the hot faucet increases the amount of hot water mixed in. But the user wants to set the temperature and flow of the water emerging from the pipe, and the controls have no simple relationship to those needs. If you want to increase the temperature, you can open up the hot

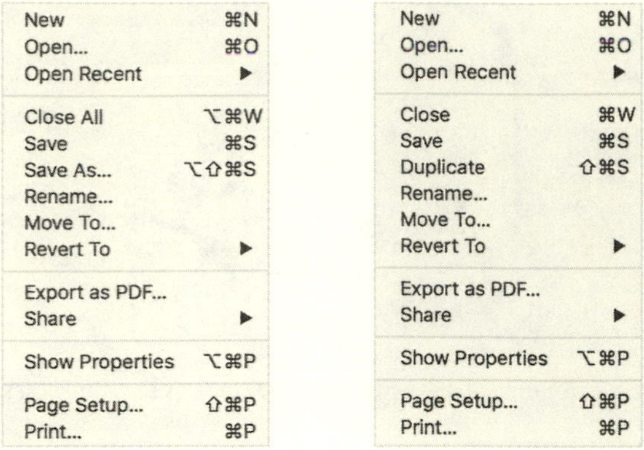

New	⌘N		New	⌘N
Open...	⌘O		Open...	⌘O
Open Recent	▶		Open Recent	▶
Close All	⌥⌘W		Close	⌘W
Save	⌘S		Save	⌘S
Save As...	⌥⇧⌘S		Duplicate	⇧⌘S
Rename...			Rename...	
Move To...			Move To...	
Revert To	▶		Revert To	▶
Export as PDF...			Export as PDF...	
Share	▶		Share	▶
Show Properties	⌥⌘P		Show Properties	⌥⌘P
Page Setup...	⇧⌘P		Page Setup...	⇧⌘P
Print...	⌘P		Print...	⌘P

FIG. 5.3 *Apple file menu: the "save as" action in the old menu (left)*
reflected the buffer concept, no longer present in the new menu (right).

faucet, but then the flow will increase, and you'll need to close the cold faucet. Likewise, if you just want to increase the flow, you need to open up both faucets, adjusting them carefully to reestablish the desired temperature. In both cases, a series of multiple adjustments is generally needed.

In the newer design (on the right), there is a single faucet that has two independent controls: rotating it adjusts the *temperature* and moving it up and down adjusts the *flow*. These controls thus have clear purposes that match the user's needs.[54]

Moving to a software example, the familiar concept of the *editor buffer* once fulfilled a need, but it's no longer a compelling one. At one time, disks were slow and the only way to make a fast text editor was to have the user edit text in a buffer in memory, periodically saving the buffer to a file. But this buffer had no apparent purpose to the user, and indeed it was confusing to non-technical users that the text in the buffer was vulnerable to being lost if the application crashed, or were closed before saving to file.

Presumably this is why Apple changed the behavior of all its applications (in os x Lion, 2011) so that changes are written to disk from the start, and "saving" to a file just entails naming the file. In other words, the purposeless *editor buffer* concept was eliminated.[55] With buffers gone, the *save-as* action (in which the contents of the buffer were saved to a new file of a given name) no longer made sense; users were now expected to duplicate the file and rename it (Figure 5.3).[56]

Andy Ostroy ✓
@AndyOstroy

Seems the only #Wall @realDonaldTrump's built is the one between him and @FLOTUS #Melania #trump

♡ 8,221   8:15 PM - May 2, 2017   ⓘ

MELANIA TRUMP liked your Tweet
Seems the only #Wall @realDonaldTrump's built is the one between him and @FLOTUS #Melania #trump pic.twitter.com/XiNd2jiLUF

FIG. 5.4 *An unflattering tweet unintentionally endorsed due to a misunderstanding over the purpose of Twitter's favorite concept.*

In all of these examples, the purposeless concepts are the result of an underlying mechanism being exposed to the user. There's nothing wrong with a text editor being implemented with a buffer. On the contrary, by applying edits first to a buffer kept in memory, and writing them out to a file on disk in the background, the editor may be able to offer much better performance. What's wrong is burdening the user with these complications.

In short, concepts, unlike internal mechanisms, are always user-facing, and must have purposes that make sense not only to the programmer but also to the user.

Concepts with Unclear Purposes: Twitter's Favorites

If the purpose of a concept is unclear to the user, it will likely be used in a way that the designer did not intend. Twitter's *favorite* concept offers a compelling example of this problem.

In May 2017, Andy Ostroy, a political analyst who writes for *HuffPost*, joked about the president's relationship with this wife in a tweet (Figure 5.4). She responded by clicking on the tweet's heart icon, apparently (and presumably unintentionally) signaling to the Twitter public that she liked the tweet. Needless to say, when she realized what had happened, she retracted her endorsement.

At issue here is Twitter's *favorite* concept. Twitter had in fact changed the visual design of the concept in 2015, replacing a star icon by a heart.[57] Apparently

FIG. 5.5 *Twitter's response to problems with the "favorite" concept: a new "bookmark" concept, accessible through the share menu (on the right), and the original concept, renamed "like," still represented by a heart icon (on the left).*

they thought this would resolve users' confusions about the concept, but it evidently didn't help this particular user. The real problem was a confusion about the very purpose of *favorite*.

Many users, it seems, thought that *favorite* offered a way to save tweets for your own reference. This was a reasonable assumption given that the term "favorite" is usually applied to a concept that has exactly this purpose. It turned out, however, that the actual purpose of *favorite* was to record your approval of a tweet for others to see—which is the purpose of the concept more commonly called *like* or *upvote*.

Twitter solved the problem in 2018 by renaming the *favorite* concept, calling it *like*, a name appropriately aligned with its purpose (Figure 5.5). To address the other purpose of allowing users to mark tweets to be saved for (private) future reference, they introduced a new concept called *bookmark*, which is accessed (confusingly) through a tweet's "share" icon (presumably on the grounds that bookmarking a tweet is sharing it with yourself?).

Exploiting Confusing Concepts: The Nanny Scam

A concept whose purpose is misunderstood is likely to be misused. The concept of *available funds* was well intentioned, but has made a rich target for scammers. When a check is deposited, some portion of the value of the check appears in the depositor's account and is immediately available for withdrawal.

In the US, this was mandated by an Act of Congress passed in 1987 designed to prevent banks from delaying the processing of deposits.

Unfortunately, many people confuse this concept with the concept of a *cleared check*, and seeing the increase in balance, believe that the check has been irrevocably validated. Criminals ruthlessly exploit this confusion. In one version known as the "nanny scam," a newly hired home help expecting an initial payment for moving expenses (of say $1,000) receives a check for a much larger amount (say $5,000) and deposits it. The employer then sends a message asking for the excess to be wired back. After the money has been sent back using the available funds, the check bounces. The deposit is retracted, effectively withdrawing the previously available funds, leaving the poor employee $4,000 in the red.

Can This Concept Really Be So Hard? The Story of Image Size

Sometimes a single concept with an unclear purpose can produce ripples of confusion. Take the concept of *image size* and the associated idea of "resolution." Even organizations that sponsor photographic contests—which you'd assume would understand these things—sometimes get this wrong, and stipulate that images must have a minimal resolution.

The problem is that the resolution of an image tells you nothing about its quality unless you also know its size. A resolution of 360 pixels per inch might seem impressive, but if the image is only the size of a postage stamp, that's not enough to get a sharp postcard-sized print.

To grasp this, you need to understand two concepts. The first is a concept I'll call *pixel array*, the ubiquitous (but once radical) idea that a picture can be represented as a two-dimensional array of colored pixels. Its purpose is image editing, and its actions include adjustments (say to contrast or lightness), which change the values of pixels, and resampling (a much more complicated action), which changes the *number* of pixels, usually to reduce the quality (by replacing several pixels with one) or to increase the quality (by interpolating additional pixels to make the image print better at larger sizes).

The second concept you need to understand is *image size*. Its purpose, which is to allow a physical size to be associated with the image, is straightforward but also strange, since we don't tend to think of digital images as having physical sizes. The image size determines the default size of the image when printed, and

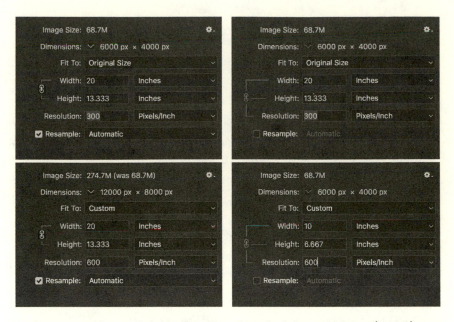

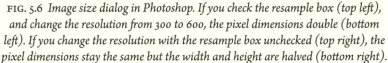

FIG. 5.6 *Image size dialog in Photoshop. If you check the resample box (top left), and change the resolution from 300 to 600, the pixel dimensions double (bottom left). If you change the resolution with the resample box unchecked (top right), the pixel dimensions stay the same but the width and height are halved (bottom right).*

its size on the page when imported into a desktop publishing app such as Adobe InDesign. In all these applications, however, the image can usually be manually scaled, so this makes *image size* a concept with a tenuous purpose.

Finally, image resolution is not a concept in its own right, but rather is a measure of printing quality assuming the image is printed at its given image size. So if the *pixel array* is 1,000 pixels square, and the *image size* is 10 inches square, the resolution is 100 pixels/inch.

If you're not already confused by this, take a look at the dialog in Photoshop (Figure 5.6) for modifying image size, dimensions and resolutions. At the top left, you can see the various parameters and confirm their relationships: at a width of 20 inches and 6000 pixels, the resolution is 300 pixels per inch. The symbol of a lock and vertical bars indicate which parameters are mutually constrained. When *Resample* is selected (the default, on the left), doubling the resolution doubles the pixel dimensions; when not selected (right), it halves the image width and height instead.

FIG. 5.7 *The tagging concept in Facebook: no mention of why you should do this.*

Even for experts, these controls are complicated and error prone. The concept of *image size* with its questionable purpose seems to be the source of all this complexity.[58]

Whose Purpose? Mine or Yours?

If you're trying to understand the purpose of a concept, a good question to start with is: *whose* purpose exactly is being served?

In social media apps, many concepts purport to be for the user's benefit, but in reality are designed to raise the company's profits, by expanding the social graph, increasing usage, or by selling more advertisements.

The *notification* concept, for example, purports to provide users with timely updates to keep them informed. But its purpose is often instead to increase "user engagement." A clear tip-off, in Facebook's case, is that while a panoply of options is available to control which events produce notifications, there is currently no option to turn notifications off completely.

The concept of *tag* likewise may seem to serve a straightforward purpose: making it easier to find posts that are about particular people. Note that when Facebook prompts you to tag someone in a photo, tellingly neither the purpose nor consequences of tagging are explained (Figure 5.7). Again, there is a hint to the real purpose if you are attentive. By default, tagging a post makes it visible not only to the friends of the person doing the tagging (as you'd reasonably

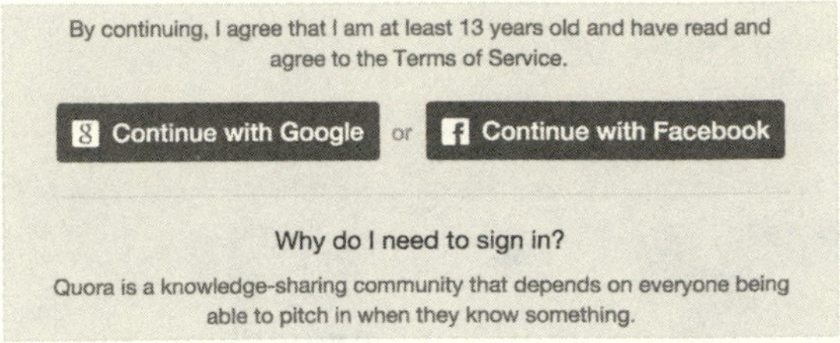

By continuing, I agree that I am at least 13 years old and have read and agree to the Terms of Service.

8 Continue with Google or **f Continue with Facebook**

Why do I need to sign in?

Quora is a knowledge-sharing community that depends on everyone being able to pitch in when they know something.

FIG. 5.8 *Quora's disingenuous explanation for why reading posts requires sign-in.*

expect) but also to all the friends of the person tagged. This neatly encourages links between two groups of friends, adding connections in the social graph.

The "Chip and PIN" security mechanism for credit cards involves two concepts with apparently distinct purposes. The *chip* concept seems designed to reduce fraud from fake cards (since it's harder to make a card with a chip inside than a card with a magnetic strip); and the *pin* concept is presumably for reducing fraud from stolen cards (since a thief will not know the PIN).

It turns out, however, that the underlying protocol is trivially broken and is susceptible to a man-in-the-middle attack. The reluctance of banks to fix (or even admit to) such problems suggests that the purpose of the *chip* concept may never have been to eliminate fraud, but rather—by giving a misleading impression that their systems are secure—to shift the blame for any fraud to consumers and retailers, and thus reduce costs to the banks themselves.[59]

Deceitful Purposes

Sometimes designers actively misrepresent the purpose of a concept, hiding a more insidious purpose:

- All question-and-answer sites have a concept of *user*, whose purpose is presumably to deter spam and low-quality answers. But many limit access so that it isn't even possible to view existing questions and answers, let alone post new ones, without first logging in. In explaining this, Quora (Figure 5.8) says: "Why do I need to sign in? Quora is a knowledge-sharing community that depends on everyone being able to pitch in when they know something." This dissembling hides purposes that are less likely to appeal to the user, such as collecting data about them, creating a more sticky experience, or pitching more focused advertisements.

Select Departing Flight:
Boston Logan, MA to Chicago (Midway), IL

FIG. 5.9 *A flight reservation app showing non-stop and direct options: note the paren-thetic explanation of the direct flight concept next to the checkbox at the bottom left.*

· A *push poll* presents itself as if it were a standard poll, whose purpose is to de-rive some useful information by aggregating responses. But instead its pur-pose is to win you over, usually for political gain, by asking you suggestive questions that are crafted to shift your viewpoint.

· The concept of *direct flight* was invented by airlines in response to early reser-vation systems that favored routes with a single flight number. By maintain-ing the same flight number across segments, "direct flights" allowed airlines to make those trips more prominent and thus more likely to be purchased. But the poor consumer misunderstanding this purpose may not realize that a direct flight is not necessarily non-stop. Nowadays, most flight aggregation sites have dropped this confusing concept, and those that use it at least add an explanation for unwary customers (Figure 5.9) and promise no change of plane, which the original concept did not guarantee.

Misfits: When Purposes Aren't Fulfilled

The essence of design is the challenge of creating a *form* for a given *context*. The desired outcome is a perfect "fit" between form and context, in the way that a piece in a toddler's wooden puzzle board fits snugly into its hole.[60]

Following this analogy, the purposes of software design describe the shape of the hole. The problem is that this shape is complex and not fully known, so it cannot be described fully or accurately. Ultimately, the only way to learn about the shape of the hole is to design a puzzle piece, attempt to insert it, and discover the *misfits*: the ways in which the piece doesn't quite fit.

Because the exact shape of the hole is unknowable, testing is essential. It is simply not possible to predict the effectiveness of a design in full until you try it in the real world. But at the same time, because of the complexity of the shape—and the fact that any given test only reveals some aspects of it—testing cannot be a panacea.[61]

You can never fully predict all the potential misfits of a design, but you can at least build on your experience of finding misfits in the past. So while a full enumeration of requirements is impossible, an enumeration of negative requirements—misfits to avoid—is feasible.

Concepts mitigate the risk of misfit in two ways. First, by breaking a design into concepts, the challenge of achieving fit for the design as a whole is reduced to a set of more manageable subproblems.

Second, concepts embody needs that occur repeatedly, and provide commonality across contexts. The misfits that are learned in one context for a concept usually apply in another. For example, a misfit of the *reservation* concept is that someone might hog multiple slots and not intend to use them all. If you're building a system that includes reservations, it helps to be reminded of this potential snag at design time, and to consider the various features that are typically used to mitigate it, such as penalizing no shows and preventing overlapping reservations (for example, of tables at different restaurants on the same night).

In the next few sections, we'll study some informative misfits. I've selected examples that illustrate the variety of misfits that can arise, and that suggest different strategies for preventing them.

A Lethal Misfit due to Poor Design

In December 2001, an American soldier in Afghanistan called in an air strike on a Taliban outpost, using a device called a PLGR (Precision Lightweight GPS Receiver, pronounced "plugger") to generate the coordinates of the target. While he was attempting to perform the calculation, the device's batteries died, so he replaced them. What he restarted the device, it seemed that the calculated coordinates were still available.

What he hadn't realized was that the device was designed to default to its own GPS location on restart. Consequently, the hapless user called in a strike

FIG. 5.10 *A* GPS *receiver (left) similar to the one that caused the accident in Afghanistan in which the operator unwittingly set the target of a bombing run to his own position, and the warning messages (right) that are now displayed.*

on his own position, and a 2,000-pound, satellite-guided bomb landed not on the Taliban outpost but on an American position, killing three soldiers and injuring 20 others.[62]

In this case, the misfit might have been predicted had the device designers considered the interactions between the *battery* concept and the *target* concept.[63] Any number of straightforward modifications to the design might have averted disaster; the one chosen and implemented in later versions of the device was to display a warning message in this scenario. (The replacement device, known as the "DAGR," is shown with the new warning messages in Figure 5.10.)

A Misfit from a Changing Context

With the advent of the pandemic, people began to give all their slide presentations online through communication apps such as Zoom, Google Hangouts and Microsoft Teams. An annoying misfit appeared. When you played your presentation, the slide presentation app would switch to full screen mode. The panels of the communication app either disappeared—leaving you disconcertingly giving a talk without knowing if you still had an audience—or obscured the slides, making it hard for you to see what you were showing.

Apple's elegant solution to this misfit was to augment the *slideshow* concept in Keynote, its slide presentation app, with a "play in window" mode, in which the slides appear in a regular window that no longer occupies the entire screen.

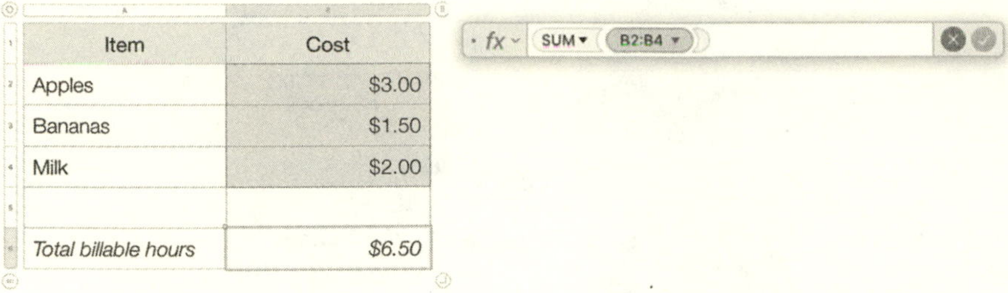

FIG. 5.11 *Defining a range in Apple Numbers: range highlighted (left) and formula (right).*

This example shows how misfits arise when the context of use evolves. A similar misfit arose in the problem I mentioned earlier (in Chapter 4) in Apple's *trash* concept—being unable to recover space on a USB key without permanently removing all the files deleted from the main drive of your computer. When the trash was designed forty years ago, personal computers did not have external drives (let alone tiny thumb drives).

An Old Misfit Returns

In a spreadsheet, a formula that computes its result from a contiguous series of cells can be expressed using the *range* concept. Instead of defining a sum of three cells as *B1 + B2 + B3*, for example, you can write something like *SUM(B1:B3)* (Figure 5.11).

The purpose of *range* is not to save typing when you enter the formula, or to make the formula more succinct when viewed—both of which could be achieved (at the linguistic level) without introducing a new concept. Rather, it's to allow the formula to accommodate the addition and deletion of cells in the series. If you add a row between rows 1 and 2, in order to include a new cell in the series, the explicit version of the formula would have to be changed manually to read *B1 + B2 + B3 + B4*, but the range formula would adjust automatically to *SUM(B1:B4)*. The operational principle of *range* may thus be stated as:

If you create a formula that depends on a range, and then update the spreadsheet by adding a row or a column within the range, the formula is automatically adjusted to include the new row or column.

The catch is how "within the range" is defined. You might think of the range as bounded by two markers, one before the first cell in the range, and one after the last cell. An addition "within" the range, therefore, could include adding a row

	Task	Time (hours)
Jan 1, 2018	Interviewing client	4
Jan 3, 2018	Making slides	5
Jan 7, 2018	Writing report	3.5
Total billable hours		*12.5*

FIG. 5.12 *A workaround for the range problem: by adding a dummy row, you can add new rows to the end of the range by inserting them above the dummy row.*

below the last row in the range (between that last row and the marker) and or adding a row above the first row.

Apple's spreadsheet app, Numbers, offers two separate actions for adding rows in general, one for below the current row and one for above (and correspondingly for adding columns). These actions are bound to keyboard shortcuts, so extending a range should be quick and easy.

Selecting the last row of a range and adding a row *below* it should include the new row in the range, but selecting the row one below the range and adding a row *above* it should exclude the new row from the range. This might sound complicated, but it's actually very intuitive: if your selection is on one of the rows in the range, then whatever action you perform, whether adding a row above or below, the new row should be in the range. But if you start outside the range, the new row should also be outside.

This in fact was exactly how Numbers *used* to behave (in the 2009 version). The current version treats the first and last row of the range differently. If you add a row above the first row or below the last row, it will not be in the range, irrespective of which row was selected and which row addition action (add above or add below) you performed.

This misfit is a major annoyance in practice. I have a spreadsheet in which I track my billing for consulting projects (Figure 5.12). Each row of the sheet corresponds to a billable period of work, with a summary row that gives the total amount of time spent. Every time I complete a piece of work, I add a row to the sheet. In the old Numbers, I would just select the row representing the last period that I completed, issue the *add row below* command, and complete the fields in the new row.

In the new Numbers, this no longer works. I can add the new row just before the last in the range, and then drag the last row up, placing it before the new row. Or I can add a spurious, empty row after the last entry, and include it in the formula (see the shaded region denoting the formula's scope in Figure 5.12). This happens to work, but only because the formula that sums the time periods happens to treat the empty cell in the dummy row as a zero. Microsoft Excel, incidentally, has exactly the same flaw (and lacks separate actions for adding rows above or below the current row).

The mystery of this misfit is not so much why Apple got it wrong, or even what it might have done to catch it (a careful consideration of the operational principle might have revealed it). More surprising is that Apple's designers knew the right design, and then apparently forgot about it. Perhaps if Apple's designers had recorded their insights in a concept catalog their best ideas would have more readily survived the transition from version to version.

Lessons & Practices

Some lessons from this chapter:

- Concept design begins by asking, for each proposed concept, the simple question: what is it for? Answering it can be hard, but pays dividends.
- For users, knowing a concept's purpose is a prerequisite to using it. Many manuals and help facilities explain details of behavior but not purposes, with unhappy consequences especially for novices.
- A concept's purpose should be cogent, need-focused, specific and evaluable. Metaphors are rarely useful for explaining what concepts are for.
- A concept without a purpose is suspect. When this happens, it's usually because the concept isn't really a concept at all, but a vestige of an internal mechanism that should not have been exposed to users.
- Confusion about the purposes of concepts leads to misuse, and can be exploited to trick users into engaging in behaviors they will regret.
- The misfits that prevent a concept from fulfilling its purpose are not easily anticipated, because the context of use evolves over time. Concepts help by giving a structure for recording experience.

And some practices you can apply now:

- If you're having trouble using a concept, first look for evidence that you may have misunderstood the concept's purpose.

- When you're explaining a concept to someone, whatever the context, start with the purpose.
- When you propose to add a concept to a product you're working on, first formulate a compelling purpose and check that it resonates with your users.
- When your team starts work on a concept, before you even make any user interface sketches, write down a terse concept description, and make sure all the designers and engineers are aligned around the purpose and operational principle.

6

Concept Composition

Until now, we've talked in detail only about concepts individually. But even the simplest app involves more than one concept, so we need to understand how concepts fit together.

In this chapter, we'll see how to combine concepts using a new kind of composition in which the actions of different concepts are tied to one another, so that when an action of one concept occurs, the associated action of another concept occurs too.

More exciting than this rather simple mechanism is what you can do with it. I'll show you a variety of compositions, starting with a trivial kind in which the concepts run largely in parallel, fulfilling their individual purposes without much interaction; then a richer kind in which the concepts are more connected and produce new functionality; and finally, the richest, in which a synergy is achieved that offers a simpler and more unified user experience than the concepts would have offered by themselves.

When you design with concepts, you can choose to synchronize them more or less tightly with each other. Tighter synchronization means more automation but also less flexibility. I'll illustrate the pitfalls of getting this wrong with some examples of both over- and under-synchronization.

Why Traditional Composition Won't Work

Software components are usually assembled in a client-and-service arrangement, in which one component plays the role of "client" and one or more others play the role of "services" provided to the client. This applies from the smallest programs (in which the client may be a function that computes the average of a list of numbers, and the services the built-in libraries that provide basic arithmetic and list manipulation) to the largest systems (in which the client may be a payroll application and the service a relational database).

This client-and-service composition allows components with more complex functionality to be built from simpler ones, and it allows components to be layered. A client only sees the components it uses in terms of the services they provide; it can't tell that those components themselves may be using other services.

For concepts, this form of composition won't work. Since concepts are by definition user-facing, we would not want one concept to be hidden behind another. Moreover, concepts must be freestanding; this is what allows them to be understood independently of one another, and reused in different contexts. In client-and-service composition, the client cannot operate independently of the services, and its behavior can't even be predicted unless the behavior of the services it uses are known.

A New Kind of Composition

Concept composition is less familiar but simpler. The concepts run, by default, independently of one another. Their actions can be invoked in any interleaving and any order, so long the individual concepts permit it.

Think of a row of snack and drink machines you might find in a railway station. You can walk up to any machine, insert a coin, and make your choice; and you can also put a coin in one machine, then put one in another machine, go back to the first machine and make a choice there, and so on. The only constraint on the order of actions is imposed by the machines themselves: that you can't get a drink before inserting a coin, for example.

To make concepts work together for joint functionality, we'll synchronize their actions. This will involve constraining both the order in which actions happen, and the relationships between the values that are passed in and out.

We could hook up the change machine with a drink machine, so that when you insert a dollar bill into the change machine and obtain four quarters in return, the four quarters are automatically used to buy a soda in the drink machine. This is a synchronization.

If we just observe the sequence of actions, each machine still behaves as before. But in composition, some sequences of actions can no longer happen—namely those in which getting change is not followed by buying a drink. Automation doesn't do new things that you could not previously have done manually; it just makes them inevitable.

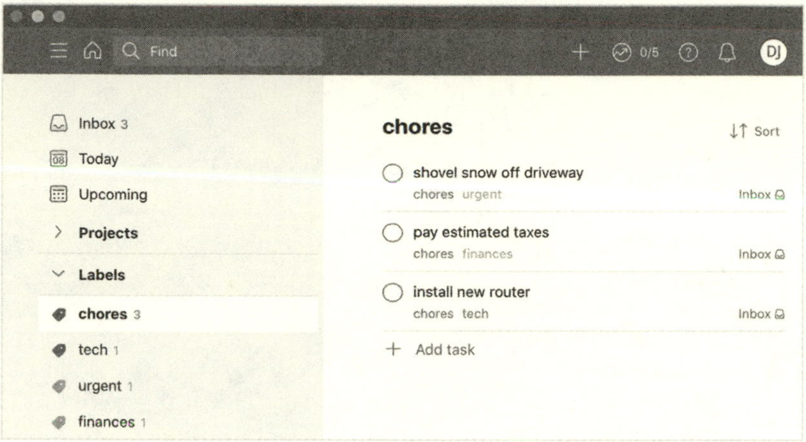

FIG. 6.1 *A todo app showing the tasks carrying the label "chores."*

Free Composition

The loosest kind of composition is *free composition*, in which the concepts are merged in a single product, but operate mostly independently of one another.

Todoist is a simple but elegant to-do app (Figure 6.1). It augments the basic functionality of to-do lists with a small number of additional features, such as organizing tasks into projects and subprojects, and attaching labels to tasks. Let's look at one of these additions, labeling, and see how it can be expressed as a concept composition.

We'll consider the *todo* concept and the *label* concept in their most rudimentary forms (Figures 6.2 and 6.3). The *todo* concept maintains a set of tasks partitioned into *done* and *pending* (these comprise the state); a task starts out as *pending* after you *add* it, and becomes *done* when you mark it as *complete* (the operational principle).

The *label* concept associates labels with items, and includes an action *find* that produces all the items that carry a given label. There's also a *clear* action that removes all the labels associated with an item. The operational principle says that if you *affix* a label to an item, and then invoke *find* with just that label (and do not *detach* the label from the item in the meantime), that item will be included in the results; and conversely, if you never *affix* a label to an item, or you detach the label, that item won't appear in the results when you query for that label.

```
1   concept todo
2   purpose keep track of tasks
3   state
4     done, pending: set Task
5   actions
6     add (t: Task)
7       when t not in done or pending
8       add t to pending
9     delete (t: Task)
10      when t in done or pending
11      remove t from done and pending
12    complete (t: Task)
13      when t in pending
14      move t from pending to done
15  operational principle
16    after add (t) until delete (t) or complete (t), t in pending
17    after complete (t) until delete (t), t in done
```

FIG. 6.2 *The todo concept defined.*

```
1   concept label [Item]
2   purpose organize items into overlapping categories
3   state
4     labels: Item -> set Label
5   actions
6     affix (i: Item, l: Label)
7       add l to the labels of i
8     detach (i: Item, l: Label)
9       remove l from the labels of i
10    find (l: Label) : set Item
11      return the items labeled with l
12    clear (i: Item)
13      remove item i with all its labels
14  operational principle
15    after affix (i, l) and no detach (i, l), i in find (l)
16    if no affix (i, l), or detach (i, l), i not in find (l)
```

FIG. 6.3 *The label concept defined.*

In practice, these concepts provide more functionality: *todo* might associate deadlines with tasks for example, and display them accordingly, and *label* might offer richer queries (on combinations of labels, for example). For understanding composition, however, these would be needless complications.

1   **app** todo-label

2   **include**

3   todo

4   label [todo.Task]

5   **sync** todo.delete (t)

6   label.clear (t)

FIG. 6.4 *A free composition of todo and label concepts. In the diagram (right), the circles on the left represent actions provided to the user, and the black arrow denotes the synchronization.*

The composition is shown in Figure 6.4. It gives a name (*todo label*) to the little "app" formed by the composition, lists the concepts that are included in the composition (*todo* and *label*), and then specifies the *synchronizations*. Note that the inclusion of the *label* concept specializes it for labeling tasks, by instantiating it with the *Task* type of the *todo* concept.

In this case, there is just one synchronization, which says simply that deleting a task in the *todo* concept causes the labels of that task to be cleared in the *label* concept. That is, when a task is deleted, its labels disappear too. Without this synchronization, a task that no longer existed from the point of view of the *todo* concept might still have labels in the *label* concept. This would result in some anomalous behaviors: if you affixed a label to a task, deleted the task, and then searched for the label, the task would appear in the results.

In case you find this underwhelming, let me reassure you that this is indeed a boring composition. We're just getting set up for some more interesting examples. What you need to understand is that putting two concepts in composition allows the user to invoke their actions in any order. Without synchronizations, the only constraints are those imposed by the concepts themselves (for example, that you can't delete a task before you've added it).

In free composition, the concepts are largely independent of one another, but there may still be some bookkeeping to do to ensure that nonsense executions are ruled out. In this case, putting the concepts together without any synchronization would allow an execution in which you *delete* a task (in the *todo* concept) and it then appears in the results of a *find* (in the *label* concept). With the synchronization, *todo.delete* is always immediately followed by *label.clear*. You can still perform the *find*, but the *clear* action will already have happened, so the *find* will not display the task that you just deleted.

83

```
 1   concept email

 2   purpose communicate with private messages

 3   state
 4       inbox: User -> set Message
 5       from, to: Message -> User
 6       content: Message -> Content

 7   actions
 8       send (by, for: User, m: Message, c: Content)
 9           when m is a fresh message not in a user's inbox
10           store c as content of m
11           store 'by' as user m is from, and 'for' as user m is to

12       receive (by: User, m: Message)
13           when m is to user 'by' and not in inbox of 'by'
14           add m to inbox of by

15       delete (m: Msg)
16           when m belongs to some user's inbox
17           remove m from that inbox
18           forget from, to and content of m

19   operational principle
20       after send (by, for, m, c) can receive (by, m),
21           and m in inbox of by and has content c
```

FIG. 6.5 *The email concept defined.*

Paradoxically perhaps, synchronization never adds new executions. It mere-ly *removes* some executions: in this case the ones that don't make sense. Every other execution that can be obtained by interleaving executions of the two con-cepts remains. For example, the composition admits a behavior comprising the following actions: *todo.add (t)*, *label.affix (t, l)*, *todo.complete (t)*, *label.find (l): t*—that is, adding a new task *t*; affixing a label *l* to it; marking the task as completed; querying for all tasks with that label and obtaining that same task *t*.[64]

Because the purpose of the synchronization in cases like this is to ensure that the same set of things exists from the point of view of each concept—that the *label* concept doesn't reference tasks that the *todo* concept doesn't know about—we might say that the concepts are "existence coupled." In every other respect, the concepts are orthogonal. The *todo* concept is oblivious to the affix-ing and detaching of labels, and the *label* concept is oblivious to whether tasks are pending or done.

Examples of this loose form of composition are common, especially in apps built on platforms that do not easily support richer synchronization between

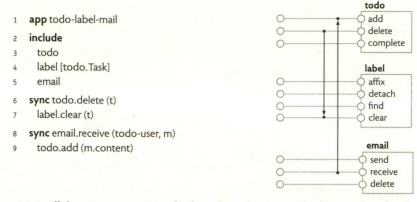

```
1   app todo-label-mail
2   include
3     todo
4     label [todo.Task]
5     email
6   sync todo.delete (t)
7     label.clear (t)
8   sync email.receive (todo-user, m)
9     todo.add (m.content)
```

FIG. 6.6 *A collaborative composition of todo and email concepts. The diagram describes the synchronization only partially: the arrow from receive to add does not imply that every email.receive leads to a task.add; as the text says, only messages to todo-user are relevant.*

their components. For example, the web services and content-management plug-ins that offer concepts such as commenting and upvoting work precisely because all they require to connect to a site is some shared identifiers (for the items being commented on or upvoted), just like the tasks in our example.

Collaborative Composition

A tighter form of composition connects the concepts together to provide new functionality that neither concept provides by itself. One of the nice features of Todoist is that you can add a task without even opening the app by sending an email message to a special email address associated with your Todoist account.

Viewed as a concept composition, this is just a synchronization of the receiving email action of the *email* concept and the task addition action of the *todo* concept. To make this a bit more concrete, let's define the *email* concept first (Figure 6.5). The state comprises a mapping from each user to the messages in that user's inbox; a record of whom each message is from and to; and the contents of the messages.

The *send* action forms a new message with some content; the *receive* action takes a message that has been previously created for the recipient, and adds it to their inbox. The operational principle expresses the idea of message transfer: after a user has sent a message with some content, the recipient can receive it, resulting in its being added to the recipient's inbox and having the content that was associated with it when it was sent.

This time, the composition (Figure 6.6) is a bit trickier. It incorporates the previous synchronization, along with a new one that ties the *receive* action of the *email* concept to the *add* action of the *todo* concept. I've given the name *todo-user* to the special email account that receives the task messages. By constraining the *receive* action to be performed by this user, we ensure that the receiving of emails by other users is unaffected by this synchronization. Note also that the *add* action binds the task being added to the content of the email message, as expected.

In the real app, the synchronization is more elaborate: you can format the subject and body of the email message to separately set the task's title and description, and you can even affix a label to the task using a hashtag notation. But the simplified synchronization that I've proposed should convey the essence of the design.

The new functionality is really just a convenience that automates a step. In principle, you could send yourself an email every time a task you wanted to add to your todo list came to mind, and then at some later point, read all those emails and add them as tasks in the todo app. The synchronization spares you this extra work.[65]

Here are some of the ways in which collaborative composition can be used:

Logging. A concept that tracks occurrences of events can be composed with other concepts. The tracking may be for diagnosis (determining why a failure happened by keeping and later analyzing the sequence of events that led up to it); performance analysis (checking the responsiveness of a service); analytics (collecting data on the users of a service and their usage patterns); intrusion detection (by searching for patterns of requests that might suggest that an attack is underway); or auditing (recording which hospital employees access health records, for example).

Suppression. In a security context, a concept may be added solely to suppress certain actions in other concepts. An *access control* concept could prevent actions by unauthorized users, by synchronizing its *grantAccess* action with the action to be authorized, so that if *grantAccess* cannot occur (as determined by *access control*), the action tied to it won't be able to occur either. The same idea can be used to limit access more generally, for example composing a *friend* concept in a social media app with a *post* concept in such a way that one user can only read another user's post if the two users are friends.[66]

Staging. The composition may bring together different stages of an activity. When making a call on a mobile phone, for example, the user may enter not the number to be called but the name of the person; a composition of the *contact* and *phone call* concepts then handles the looking up of the number and the placing of the call respectively. This kind of separation into stages allows features such as *call forwarding* to be treated as concepts. A similar pattern arises in the browser request pipeline, which starts with a lookup using the *domain name* concept that translates a domain name to an IP address, which is then used by the *http* concept for the request itself.

Notification. Most apps and services offer notifications to users: calendars send reminders; help forums let you register to be notified of replies to a question you asked; online stores send acknowledgments of purchases; shipping companies send delivery updates; social media apps tell you when your friends have posted new content. All these can be done with a *notification* concept that offers an event tracking action that is synchronized with the actions of interest in the other concepts, and notification actions that follow spontaneously (and whose timing, medium, and frequency can often be configured by the user).

Mitigation. Sometimes, free composition gives too much latitude to users, leading to undesirable behaviors that can be mitigated by a collaborative composition. Many social media platforms, for example, compose a *post* concept with an *upvote* concept that lets users rate individual posts. If the *post* concept allows editing, this creates a dilemma, because a user might receive lots of positive upvotes for a post only to then change its content entirely, giving the misleading impression that the new content received all those approvals. One common fix for this (used in Slack, for example) is to add an indelible marker to posts that have been edited. Another is to synchronize the *edit* action of the *post* concept with an action in another concept that undoes some of the approval. In YouTube, for example, a user can *pin* a favorable comment to a video of theirs. But if the comment is edited, it will be automatically unpinned, by a synchronization between the *comment.edit* and *pinning.unpin* actions.

Inference. Sometimes a user's actions are not executed directly, but are inferred from other actions. Most communication apps distinguish between read and *unread* items, and allow users to toggle their status. But to mark an item as read for the first time, an app typically synchronizes another action (such as opening the item, or scrolling through it) with the action that marks it as read.

Bridging separated concerns. By separating concerns with free composition, concepts usually improve the clarity and usability of an app. On a mobile phone, for example, the *cellular* and *wifi* concepts let you manage your use of cellular data and local networks independently of each other, and of the apps that use the data. Sometimes, though, the need arises to couple concepts that have been separated. Apple's Podcasts app, for example, gives you the option of preventing a *podcast* from being downloaded using the cellular connection (and thus consuming data against your quota when you might instead have obtained it free through Wi-Fi).[67]

Synergistic Composition

In free composition, a software product is assembled from largely orthogonal concepts, each bringing its own functionality, with synchronization being used only for bookkeeping. In collaborative composition, synchronization creates connections between concepts that provide automation, and thus some new functionality that the individual concepts do not provide by themselves.

In synergistic composition, something more subtle occurs. By synchronizing the concepts even more tightly, the functionality of one concept comes to enhance another concept's fulfillment of its own purpose. The overall value of the composition is now more than the sum of the values of the concepts.

To illustrate this phenomenon, suppose that, in the composition of the *todo* and *label* concepts, we were to represent the pending/done status of tasks with a built-in label *pending* that is affixed automatically when a task is added and detached when the task is completed. This can be described with two synchronizations (Figure 6.7): one to affix the label when a task is added, and one to detach it when the task is marked as complete. For consistency, I've added a third synchronization that causes the task to be marked as complete when the label is affixed.

The advantage of this composition is that the label query functionality now incorporates whether tasks are pending or not. There's a single uniform interface, and in a richer version of the *label* concept that offers a logical query language, you could ask for tasks that are "pending and urgent," for example. Moreover, the state component of the *todo* concept that remembers whether a task is pending or done is no longer necessary, since that information is now stored in labels.

1 **app** todo-label-syn

2 **include**

3   todo

4   label [todo.Task]

5 **sync** todo.delete (t)

6   label.clear (t)

7 **sync** todo.add (t)

8   label.affix (t, 'pending')

9 **sync** todo.complete (t)

10   label.detach (t, 'pending')

11 **sync** label.detach (t, 'pending')

12   todo.complete (t)

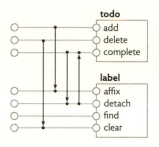

FIG. 6.7 *A synergistic composition of todo and label concepts.*

This simplified setting can only hint at the benefits that this kind of composition can bring. In the next section, we'll study an example that's more powerful but also surprisingly complicated, showing how subtle synergistic composition can be. Before that, some other examples of synergistic composition:

Gmail labels and trash. Google's email app, Gmail, uses labels synergistically in exactly the way I just described. When an email message is sent, the label *sent* is affixed to it automatically, and the button labeled "sent" that opens a list of sent messages is simply bound to a query for messages carrying the *sent* label. The same idea is extended to the *trash*: deleting a message affixes the *deleted* label to it, and removing that label restores the message.[68]

Moira lists and groups. MIT uses a system developed internally in the 1980s called Moira for managing mailing lists. To let multiple users maintain a mailing list, you can create a second list of those users; that list can then be assigned as the owner of the first list. To grant or withhold control, you can simply add or delete a user from the second list. This is a lovely synergy in which the *mailing list* concept is composed so thoroughly with an *administrative group* concept that the latter requires no interface of its own.[69]

Free samples and shopping carts. Some online stores include in their shopping carts items such as a free samples (or catalogs, etc.) that have *not* been purchased by the user. This is a synergistic composition between the *shopping cart* concept and the *free sample* concept; the action that adds a free sample to the order is synchronized with the action that adds items to the shopping cart. This is good for the user (who sees all their items, including the free ones, in one place), and it makes things easier for the developer (eliminating the need to store the free samples separately). But, as with many synergies, unexpected snags can arise.[70]

Photoshop channels, masks and selections. The preeminent example of synergy appears in Adobe Photoshop, in which the *mask*, *selection*, and *channel* concepts work together with remarkable power.[71]

The Beautiful Synergy of Trash & Folder

You may have been surprised that when I introduced the *trash* concept in Chapter 4, I treated the trash as just a set of items. In the most familiar instances of the concept (on the Macintosh or Windows desktops), the trash is not a set of items but a *folder*.

In some early versions of the concept, the trash *was* just a set, and if you deleted a folder, its contents were disaggregated and placed individually in the trash. Obviously this made restoring folders difficult, and the better design (which Apple had used from the start) fused together the two concepts.[72]

The modern design can be understood as a very skillful composition of these two distinct concepts, *trash* and *folder*. By viewing it in these terms, we can separate out the various aspects of behavior. To understand the essential idea of the trash—that items can be deleted and later restored, or permanently removed by emptying—you only need to know about the *trash* concept; to understand how the items in the trash appear, you only need to know about *folder*.

The synergy this creates is evident in the parsimony of the user interface. No special actions are needed to list the items in the trash, since in this respect, it's just a regular folder, to which you can apply sorting, searching and so on in the regular way. Nor is a special control needed for restoring an item; you just move it out of the trash folder. And of course, the synergy is what allows the trash to retain the structure of a deleted folder so that it can be restored in one piece.

The synchronization that makes this possible is not complicated. Moving a file to the trash (in the *folder* concept) is synchronized with deleting the file (in *trash*); moving a folder is synchronized with deleting all the folders and files it contains; and likewise, moving a folder out of the trash is synchronized with restoring all those folders and files.

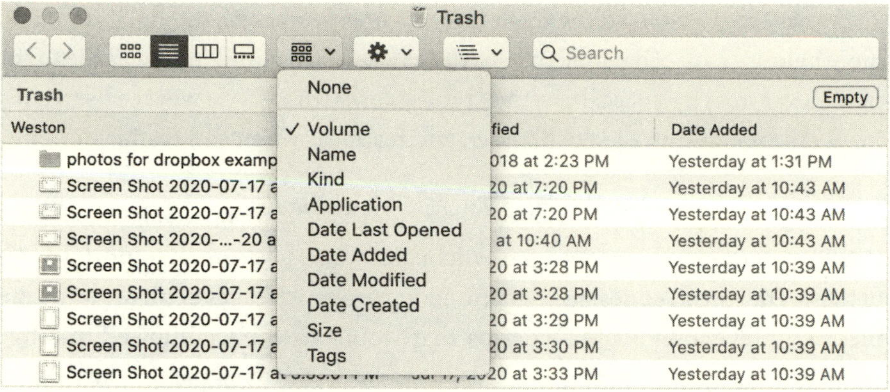

FIG. 6.8 *Synergistic composition of the trash and folder concepts. Items can be sorted by "date added," which corresponds to the date of deletion, elegantly reusing a general feature of folders. The sorting into volumes is a more troubled feature, since it applies only to the trash folder.*

Synergies Are Rarely Perfect

Perfect merging of functionality across concepts is rarely possible, and so most synergies have some costs. The trash is not quite a folder like any other; most obviously, it needs to offer the *empty* action, hence the little button that appears in that folder alone.

The designers of the Macintosh trash have tried their best to minimize such non-uniformities. They resisted, for example, including a "date deleted" field, which would have only applied to the trash folder, but cleverly included a "date added" field, which works for all folders, and for the trash happens to allow sorting by deletion date.

More troublingly, in the Macintosh desktop, there is only *one* trash folder, even if there are multiple drives. Unlike any other folder therefore, the trash can "hold" items from different volumes; to make this clear, recent versions of macos allow the items in the trash to be grouped by volume, a feature not available for other folders (Figure 6.8).

Representing the trash as a folder can be confusing at times: hence the uncertainty in my saying that the trash can "hold" items. In Chapter 5, I mentioned the scenario in which you insert a removable drive into your laptop, and move some items from it to the trash folder in the hope of making space on the drive. Since there is only one trash folder, and it "belongs" to your laptop, you might imagine that this action would free up space on the drive.

But, as the division of the trash into volumes indicates, the trash folder no more belongs to the machine than to the external drive, and moving a file to the trash never frees up space. If you eject the external drive, you will see its trashed items disappear from the trash folder, and reappear when you reattach it.[73]

Synchronizing Too Much and Too Little

When you compose concepts, the synchronizations become an important part of the overall product design. Synchronizing too much takes control away from the user, preventing some scenarios that would have been allowed in a free composition. Synchronizing too little, conversely, burdens the user with work that might have been automated. It can also admit unexpected and undesirable behaviors, sometimes with catastrophic consequences.

Over-Synchronization & The Strange Case of The Canceled Seminars

In Apple's Calendar app, the *calendar event* concept that supports the storing of events at given times is composed with an *invitation* concept so that one user can send another a tentative event, which the second user can accept or decline.

In the original design, deleting an invitation event presented users with a quandary. The *delete* action was unhelpfully coupled with the *decline* action, so you could not delete the event without notifying the issuer of the invitation that you were declining it. This was bad enough if you simply wanted to clear space in your calendar without offending a friend. It was far worse if the invitation were spam. In that case, the notification would aid the spammer, confirming your email address as valid and increasing the likelihood of future spam!

For many years, a clunky workaround was the only way out: to create a new calendar, move the event to it, and then delete the new calendar in its entirety. Eventually (sometime in 2017), Apple uncoupled the delete and notify actions (Figure 6.9).

The very same design flaw, but in Google Calendar, turned out to be the source of a puzzling problem that my lab had with seminar announcements. Often a seminar cancellation would be sent shortly after the initial announcement. The organizer would then send a follow-up, reassuring people that the seminar had not in fact been canceled. It turned out that the cancellations were spurious. Individuals in the lab were receiving notifications of research seminars, and adding them as events in their personal calendars. Then when

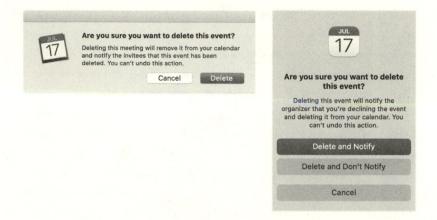

FIG. 6.9 *The original Apple Calendar dialog that unhelpfully always synchronized deleting an event with notifying the sender (left), and the most recent version (right) that fixes the problem by making the synchronization optional.*

someone deleted such an event, a cancellation message was sent spontaneously by Google Calendar to the email address associated with the initial invitation—which in this case was a listserv with more than a thousand members!

Some other examples of over-synchronization:

Tumblr's questionable design. In the Tumblr blogging platform, if you wanted to allow people to comment on your post you could insert a question mark at the end of the post's title. In this kind of synchronization, whether one action (here, creating a *post*) produces an accompanying action (enabling *comments*) depends on the exact *content* of the arguments to the first action. This is not only an unwelcome synchronization—what if your title were a rhetorical question?—but also makes the *title* concept no longer generic (see "Concepts are generic" in Note 48) and thus harder to understand. Later, Tumblr made it possible to check a box instead.

Replies in Twitter. A similar issue arose in Twitter. Until mid-2016, a *tweet* that began with a username was interpreted to be a reply. This led, amongst other things, to a convention in which people wanting to mention someone else at the start of a tweet, but not to make it a reply, would insert a period before the name ("*.@daniel Really?*").

An unwanted Google synchronization. If you have a Google account whose *username* is an external *email address*, and you add Gmail functionality to the account, your username will be automatically changed to match the new Gmail email address, and you will not be able to reinstate the old username and email address![74]

Epson's tyrannical printer driver. In its photo printers, Epson understandably wants to prevent users from making certain combinations of settings that might damage the printer. For example, it seems reasonable that you should not be able to select "thick" for the *paper option* along with a *paper source* setting that feeds paper from the top (which requires it to bend as it enters the printer). But the printer driver goes further, and prevents you, when selecting the top feed, from using most of the fine art paper settings, presumably on the grounds that these settings only apply to thick papers. This is wrong, however; many fine art papers are thin and flexible, and this constraint forces you either to feed such papers through the front (which is much less convenient since each sheet must be manually loaded), or to feed them through the top, but to print with incorrect ink settings.

Under-synchronization and a Group That Could Never Be Joined

A year or so ago, I wanted to create an online forum for neighborhood discussions, so I set up a Google Group for the purpose. Since I did not know in advance the user names of my neighbors, I sent them a link to the group, suggesting that they request to join.

Unfortunately, this did not work, and people were unable to even access the page with the ask-to-join button. I thought I'd set the group up correctly; in particular, under "permissions" (Figure 6.10, top) in response to "Select who can join," I'd chosen "Anyone can ask."

It turned out there was a different setting that determined whether the group appeared in a directory of groups (Figure 6.10, bottom). Unless the visibility of this feature was set to "Anyone on the web," the group was not only excluded from the directory but was not accessible at all, even for join requests!

The lack of a synchronization between the action that determines who can join (in the *permission* concept) and the action that sets visibility (in the *group directory* concept) got me into this mess. Since then, Google has tweaked the design. Both controls are placed on the same page (Figure 6.11), but they're still not synchronized, so selecting "Anyone can join" for "Who can join?" won't work unless you also change "Who can see group?" from the default setting to select "Anyone on the web."

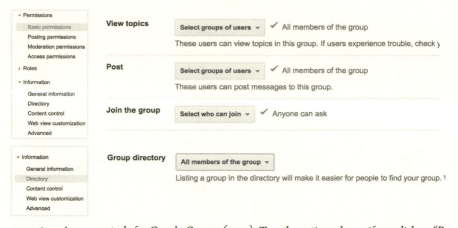

FIG. 6.10 *Access controls for Google Groups (2019). Top, the options shown if you click on "Basic Permissions" in the "Permissions" tab; bottom, the options shown if you click on "Directory" in the "Information" tab. With the default setting of "Directory," the permission setting was ineffectual, since the page displaying the ask-to-join button was private to members!*

Some other examples of under-synchronization:

Lightroom import. Adobe's photo cataloging and editing app, Lightroom Classic, offers a rich importing mechanism that not only copies photos from a card or camera to your hard drive, but can perform an array of additional tasks, such as: moving the copied photos to a preferred location; adding copyright information to the metadata; applying development settings to every photo; adding keywords; building previews; and ejecting the external drive or flash card holding the originals. All of these synchronizations were optional, and were controlled by preferences in a fairly complicated user interface dialog.

In 2015, the Adobe team released version 6.2 of Lightroom, which included a simplification of the import dialog, removing some of the synchronizations that expert photographers had come to rely on. The reaction from users was so fast and so negative that Adobe actually reverted the update.[75]

Google Forms, Sheets, and data visualization. In a lovely example of synergy, Google Forms use Google Sheets, Google's spreadsheet app, as the repository for data collected. The Forms app also includes a nice visualizer that generates pie charts, histograms, and so on to summarize the data. Unfortunately, the visualization is synchronized with a *different* copy of the data, distinct from the sheet. As a result, edits to the sheet (for example, data cleaning such as removing duplicate form submissions) are *not* reflected in the visualization.[76]

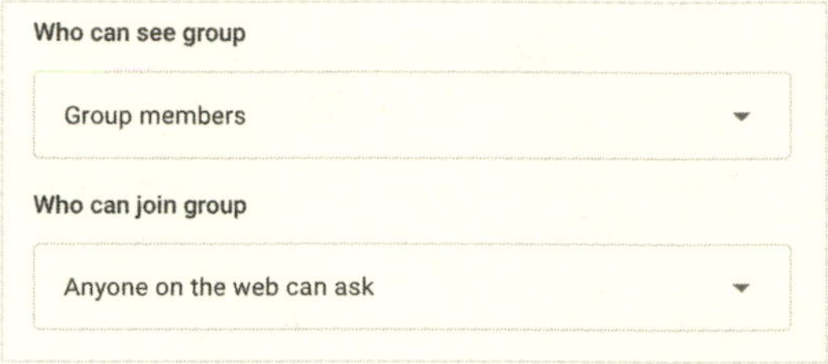

FIG. 6.11 *A more recent version of Google Group permissions (February 2021), which helpfully consolidates the permissions for viewing into a single setting, and places it on the same page as the permissions for joining. Unfortunately the default is as shown, and asking to join requires being able to see the group.*

Zoom's sticky hands. In the Zoom video-conferencing app, a participant can raise a virtual hand to signal to the host a desire to speak. While not talking, participants typically mute their microphones to reduce background noise. Then when called on, a participant unmutes, makes a point, and mutes again. But most participants forget their raised hands, confusing the host who, noticing a raised hand later, is uncertain whether that person just forgot to lower it, or wants to speak again. Synchronizing the *raised hand* and *audio mute* concepts might eliminate this annoyance.[77]

Therac 25 radiotherapy machine. A series of catastrophic accidents with a radiotherapy machine in the late 1980s was eventually traced to a synchronization flaw. The source of the radiation was an adjustable electron beam that through either (a) a magnetic collimator to focus the electrons, or (b) a flattening filter that converted the electrons to X-ray photons (depending on the position of a rotating turntable).

Patients were irradiated in one of two modes: either directly with electrons or with X-rays. The X-ray production required a high electron current—far more than could ever safely be administered directly. The intent, therefore, was to ensure that the flattening filter was in place whenever the electron beam current was high. Tragically, a flaw in the synchronization mechanism sometimes resulted in the high current being delivered directly to patients, resulting in massive, fatal overdoses. Although the flaw has been attributed to a programming error—a bug in the code that failed to ensure the intended synchronization—a better design might have prevented it.[78]

Lessons & Practices

Some lessons from this chapter:

· Concepts are not composed like programs, with larger concepts subsuming smaller ones. Instead, each concept is exposed to the user on equal terms, and a software app or system is a collection of concepts running in tandem.

· Concepts are composed by synchronizing their actions. This never adds new concept executions, but constrains existing ones, eliminating some sequences of actions that would have been possible for a concept in isolation.

· In free composition, concepts operate independently of one another, constrained only by some bookkeeping to ensure (for example) that concepts have a consistent view of which things exist.

· In collaborative composition, concepts work together to provide new functionality through automation.

· In synergistic composition, the concepts are yet more tightly intertwined, with one concept's functionality helping another concept fulfill its purpose.

· Composition offers an opportunity for creative design even when the concepts themselves are familiar. Synergy is often the very essence of a design, bringing unexpected power from the combination of simple parts.

· Synchronization is an essential part of software design. Not enough synchronization can lead to inappropriate or confusing behaviors, and miss opportunities for automation; too much can limit the user's options.

And some practices you can apply now:

· If a concept you're designing seems complicated, try thinking of it instead as a composition of simpler concepts, which may be easier to describe and justify (by having a clearer purpose and more compelling operational principle). (More on this in Chapter 9.)

· When picking concepts, be on the lookout for familiar concepts to reuse (Chapter 10). For example, you might identify a *notification* concept and then see ways to use notifications more uniformly throughout your app.

· In design, decide first which concepts to include, and then how they to synchronize them. Start with synchronizations to remove bad behaviors. Then consider automations, but make sure to leave enough flexibility to the user.

· Look for synergies in which you can simplify a concept by composing it with another. But remember that perfect synergy is rarely attainable.

7

Concept Dependence

When concepts are composed, they play specific roles in their relationships to one another. For example, when we bring together *label* and *todo* so the user can attach labels to todo tasks, the *todo* concept becomes a kind of subject to which the *label* concept is applied.

The composition itself is symmetrical, because synchronization treats synchronized actions as equals. Nevertheless, composition can introduce an asymmetry in the way in which one concept augments the functionality of another. The *label* concept extends the functionality of the *todo* concept: now, in addition to adding tasks, we can label them. The converse would make no sense: nobody would start by building an app for labeling things, and then extend it with tasks that can be labeled.

These asymmetries reveal important structure in a software product, which this chapter explores. I'll explain a *dependence* notion amongst concepts that can be depicted in a simple diagram, which offers a helpful summary of the concepts and their roles.

It may seem paradoxical to introduce a notion of dependence when I have insisted that concepts are mutually independent. There is in fact no paradox here. The very essence of concepts is indeed that they can be understood and implemented in isolation. The dependencies discussed in this chapter arise from the role that concepts play in the context of the product as a whole, and is more a property of the product than of the concepts it contains.

Growing a Software Product Concept by Concept

Some software products must emerge from the womb fully grown. An aircraft or nuclear power station cannot be deployed with software that is only a "minimum viable product," with a plan to adjust the software as needs arise.

But in most cases an incremental development is better, because it allows the developers to get early feedback on their design work, assess the value of the work already deployed, and handle misfits as they are discovered. So it's useful to think about designing a new software product as *growing* it, a few concepts at a time.

Not all growth involves the addition of concepts. Sometimes a concept will be removed, perhaps because it is found to be less useful than expected, or because it has some critical flaw that is not easily remedied, or because its functionality can be subsumed by extending another concept. Sometimes the existing concepts will be refined and polished, ideally becoming not only more powerful but also more compelling (in their purposes and operational principles). And sometimes—the most exciting case—a synergy is discovered (with or without the addition of a concept) that extends the product's capabilities without a concomitant increase in complexity.

Unbridled growth can be the downfall of excellent products. Ambitious redesigns of small but successful systems are especially subject to what Fred Brooks, an influential IBM manager, called "the second-system effect" in which overconfidence leads to bloated and needlessly complex solutions.[79]

For all these reasons, it's useful to be able to represent succinctly the possible growth of a software product, and—equally importantly—the ways in which it may be trimmed back. This is what the concept dependence diagram provides.

Building the Concept Inventory

I mentioned in Chapter 3 how concepts can give you a map of an application—a kind of concept inventory that gives you an overview of its functionality and purposes. Let's see how this map comes about, using an imaginary app as an illustration.

Throughout, I'll discuss it as if you were designing the app alone; of course, in practice, most design work is done in teams. I'll also talk only about which concepts to include, ignoring all the critical questions about the actual design of the concepts.

Suppose you love bird song, and want to build an app to help people identify birds from their calls. That's the essential need: someone hears a bird call and wants to know what kind of bird it came from. And presumably people who want this also want to hear the song of a particular bird species.

You think about how such an app might work. Maybe a user uploads an audio track, then other people listen to it and make suggestions. Now you start brainstorming some concepts. Before inventing a new one, you try some existing ones for fit. How about the *q&a* concept used in forums (such as StackExchange, Quora, Piazza, and so on), in which someone asks a question and other people provide answers?

In fact, you might ask, why not just use one of those existing apps? Often that will be the best approach to solving a problem. To make sure that it isn't, you'll want to note some limitations of the existing solutions and check that they really matter. Perhaps in this case you discover that none of them let you easily upload and playback audio recordings; maybe it's important to have a nice integration so you can record a song and post it in a few clicks.

So now let's suppose you've convinced yourself you need to design a new app, and you have some *seed concepts*, such as *q&a* and *recording*. What additional concepts are necessary to make a coherent app? Obviously, since you're relying on crowdsourcing, you'll want some concept for achieving consensus, so you might add *upvote* so that users can approve an answer.

As you brainstorm how your app will be used, you realize that you'll need to somehow consolidate the bird identifications. Users will likely want to search for particular species, and listen to recordings for which there are confirmed matches. So you tentatively add an *identification* concept, unsure of exactly how it will work. Perhaps when a user answers a question, they can propose an identification by inserting a hashtag, and your app will automatically extract links between species and recordings from the upvoting of answers.[80] Finally, you decide to add *user* to authenticate contributions. At this point, you have the rough outline of an app: BirdSong 0.1.

An Inventory of Generic Concepts

The concepts we have so far, along with their purposes, are:
- *q&a*: support community response to questions
- *recording*: allow upload of audio files
- *upvote*: rank contributions based on individual (dis)approvals
- *identification*: support crowdsourced assignment of objects to categories
- *user*: authenticate content and actions

Note that each concept has been given a generic purpose. Of course, in the context of the articular app, each concept will be specialized. For BirdSong 0.1, the questions will be about bird calls; the audio files will contain bird songs;

the contributions that are upvoted will be proposed answers or identifications. Even the *identification* concept, which might have seemed to be very bird-specific, has been cast in more general terms, in the hope that this will make it easier to draw inspiration and lessons from related concepts (such as *tag* in Facebook).

Making the concepts generic not only makes the reuse of design knowledge from previous applications possible. It also helps simplify the design. The less bird-specific any concept is, the easier it will be to understand. Maybe you're tempted, for example, to include in the *identification* concept whether the bird is male or female. A priori, however, this is a bad idea: until you have more experience and understanding of the app and how it will be used, there's no reason to believe that including this distinction (rather than just treating the male and female as separate birds) is any more important than any other. If you wanted to associate different birds, a more plausible start would be to enrich the *identification* concept with a notion of birds being *related*; this could then accommodate not only distinctions of gender, but other variants within a species.

The Concept Dependence Diagram

Because each concept is generic and freestanding, there are no concept-to-concept dependencies in the traditional software engineering sense. But there is a different kind of dependence between concepts, related not to the concepts themselves but to their role in the application as a whole.[81]

Take the *upvote* concept. Clearly, there's no point having this concept in the app unless there is something to upvote! Presumably, what's being upvoted is an answer to a question ("That's a #sparrow singing").[82]

So we'll say that *upvote* "depends" on *q&a*, because if the *q&a* concept were missing, there would be no reason to include the *upvote* concept. Collecting together all such dependencies gives the diagram of Figure 7.1.

Sometimes a concept's presence may be justified by any of several other concepts. In that case, we'll mark one of the dependencies as the primary dependence (with a solid line), and the others as secondary (with dotted lines). A secondary dependence represents additional, but less compelling, reasons to include a concept.

Thus *user* has a primary dependence on *q&a*, because the main reason for including user authentication is to ensure that questions and answers can be

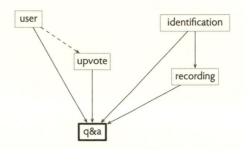

FIG. 7.1 *Concept dependencies for a bird song app. A solid arrow denotes a primary dependence and a dashed arrow a secondary dependence. The core concepts are in bold.*

reliably associated with individuals; and a secondary dependence on *upvote*, because the authentication could be used to prevent double voting too. The second usage is less essential; we could use IP addresses or browser IDs for that purpose instead.

The diagram tells us which concepts are core to the app and which could be omitted. Because every concept depends directly or indirectly on *q&a*, an app can't exist without it—if it includes any of these concepts, it must include this one. But it could exist alone, with no other concept to augment the functionality. This would admittedly be a rather feeble app: lacking *user* means no authentication; lacking *upvote* means no crowdsourcing; and lacking *recording* means that questions would have to describe songs in words, or perhaps link to files elsewhere on the web.

Any subset of the concepts can comprise a consistent app, so long as there are no dependence edges pointing out of that subset. So *q&a*, *recording* and *upvote*, for example, make an app. In contrast, *identification* and *q&a* do not make an app, because *identification* depends on *recording*. In the context of our app, the *identification* concept provides reverse lookup: given a particular identification, it leads the user to the relevant bird songs. Without *recording*, this role cannot be fulfilled.[83]

The diagram therefore describes not just one app, but a whole family comprising all the apps that could be built from these particular concepts—what software developers would call a "product line." Each consistent subset represents an app that might be built.

The subsets can also represent stages of development. At any point in the development, you'd like to have a consistent subset implemented in order to be

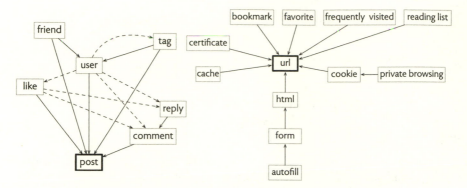

FIG. 7.2 *Concept dependencies for Facebook (left) and Apple Safari (right).*

able to evaluate it as a coherent unit. If you implemented a set that included *up-vote* but not *q&a* you'd have trouble making a compelling demo or test, because there would be nothing to upvote.

Finally, the diagram provides an *explanation order*. You can't explain an app all at once, so you explain the concepts in order, one or two at a time. But what orders make sense? Dependencies tell us how to avoid introducing a concept before it can be motivated. So if we were explaining our bird song app to a novice user, this order would make sense

q&a, upvote, user, recording, identification

but this would not

upvote, q&a, user, identification, recording

because it introduces *upvote* before *q&a*, and you can't explain upvoting without something to upvote (just as you can't demo it).

The Structure of Some Familiar Apps

To illustrate the concept dependence diagram further, and how it can provide insight into designs, let's look at the structure of some familiar apps.

Facebook. Figure 7.2 (left) shows the key concepts of Facebook and their relationships. The base concept, of course, is *post*. Comments are on posts, so the *comment* concept depends on the *post* concept. The *reply* concept offers threaded conversations about comments. The *user* concept offers authentication for posts primarily, but also for comments, replies, tags and likes.

The *friend* concept is interesting; since its purpose is to allow users to control access to their posts, it depends not only on *user* but also on *post*. The *tag* concept

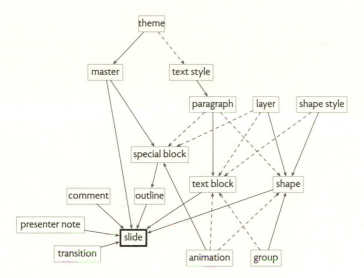

FIG. 7.3 *Concept dependencies for Apple Keynote.*

involves identifying users that appear in posts, so it depends on *user* and *post*. Finally, the *like* concept depends primarily on *post*, but is also used for comments and replies.[84]

Safari. Figure 7.2 (right) shows the key concepts of Apple's Safari browser. As you'd expect, *url* is the base concept; for easier layout, I've put it in the middle rather than at the bottom. The *url* concept embodies the idea that resources (namely web pages) can be obtained by sending requests to servers with persistent names (namely, uniform resource locators). The *html* concept allows these resources to be marked-up pages, but most of the browser concepts are not dependent on this and could still be used in a browser (admittedly, a rather feeble one) that does not include HTML rendering. The *cache* concept depends only on the *url* concept; it helps the browser run faster by storing the resources returned by previous requests to a given URL. The *certificate* concept makes sure that the server the browser talks to really corresponds to the domain name in the URL, and so depends only on the *url* concept. The *private browsing* concept offers a mode in which cookies are not sent to servers, safeguarding the user's identity, so it depends on *cookie*.

At the top of the Safari diagram is the *bookmark* concept and three variants of it: *favorite*, which, like *bookmark*, allows you to save URLs to visit later, but displays these URLs in the toolbar and on every new tab you open; *frequently visited* which creates bookmarks spontaneously from sites you've visited many times; and *reading list*, which is also just like a bookmark, but tracks whether a page has been read (and also downloads the page for offline reading). The proliferation of these very similar concepts, and the subtle differences between them, suggests an opportunity for more synergistic design. The ability to access pages offline and mark pages as read could be added to all bookmarks. And frequently visited sites, like favorites, could be added as a special folder within the regular bookmarks so that they could be deleted if unwanted.[85]

Keynote. Figure 7.3 shows the key concepts of Apple's slide presentation app, Keynote. As you'd expect, the *slide* concept sits at the base. The *special block* concept generalizes the title, body, and slide number which appear optionally on each slide, and are given default formats in *master* slides. The *theme* concept allows a collection of masters to be shared across presentations (for consistency and ease of use), and naturally augments the *text style* concept (by playing the role of a *stylesheet* concept that brings together a collection of styles for reuse across documents).

In addition to the *special block* concept, there is a separate concept of a *text block*, as well as a *shape* (which can also hold text, but does not expand automatically to fit the content). Text is always broken up by *paragraph*. There are two instances of the standard *style* concept, one for text in paragraphs and one for shapes. The *layer* concept supports stacking of shapes and text blocks (with the "send to back" and "bring to front" actions). The *animation* concept primarily supports progressive reveal of points in special blocks, but can also sequence the appearance of shapes and text blocks.[86]

Lessons & Practices

Some lessons from this chapter:

- Concepts are freestanding, and independent of one another: a concept can be understood, designed and implemented by itself. This independence is key to the simplicity and reusability of concepts.
- In the context of a software product, dependencies arise—not that one concept relies on another for its correct operation, but because inclusion of one concept may make sense only if another is present.

· The dependence diagram gives a succinct summary of a product's concepts and the motivations for including them. It helps plan the order of design and construction, identify subsets, and structure explanations.

And some practices you can apply now:

· When you are designing an app, think about growing it one or two concepts at a time. At the start, identify a few seed concepts that will form the basis for all subsequent growth.

· Draw a dependence diagram to get a succinct view of the concepts in your app and their relationships. Each time you add a concept to your design, think carefully about which concepts it depends on: more is generally better, because it means the concept is used more extensively.

· When considering what order to prototype or build the concepts in, refer to the dependence diagram, so that at any point you have a coherent subset.

· To explore ways in which an app might be simplified, evaluate the consistent subsets and estimate how much value each brings. Perhaps there is a subset that brings most of the value for only a small part of the cost.

· When writing a user manual or developing training materials, use the ordering defined by the dependence diagram to present concepts in the most efficient and rational sequence.

8

Concept Mapping

You can think of the concepts of an app as running in the background, behind the user interface. The interface provides buttons to activate the actions of the concepts, and visualizations of the concept states. So when a user clicks on the "like" button of a social media post, an action of the form *upvote.like(u,p)* is activated, telling the *upvote* concept through its *like* action that user *u* approves of post *p*. The result of this action—a change in the state of the *upvote* concept—is then reflected in the updated count of likes displayed to the user.

Creating a user interface involves more than visual design; its essence is the design of a *mapping* from the underlying concepts to their material form in the interface. The interface designer shapes this mapping, usually by creating multiple screens and dialogs, connected by flows and links, and then embedding within them controls and views that connect to the concepts actions and states.

The design of mappings has been extensively studied by human-computer interaction researchers, and the guidelines that have been developed—mostly at the physical and linguistic levels of design—apply equally well to systems that have been designed with concepts.[37]

But concepts offer a chance to refine (and maybe even rethink) the relationship between the conceptual level of design and the physical and linguistic levels. So this chapter focuses on examples that illustrate how tricky and intricate mapping design can be, and how deeply it must be informed by the underlying concepts.

How to Make a Simple Concept Hard

Even if the underlying concept is simple, it's still possible to design a mapping that makes it hard to use. Last week, a message popped up on my desktop asking me if I wanted to upgrade to the latest version of Oracle Java. I clicked yes. When I ran the installer, it displayed a dialog with two buttons, one labeled "Install" and the other "Remove" (Figure 8.1).

FIG. 8.1 *A puzzling dialog in a Java installation process. What does "remove" mean?*

Now you might imagine there was no room for confusion here, and yet I still managed to confuse myself. Presumably the install button invokes the action that installs the new version of the software. But what does the remove button do? Apparently, it removes whatever version I have currently installed, and then does nothing.

These might seem like reasonable interpretations; after all, that's pretty much what the words "install" and "remove" mean. So why was I confused? Well, for one, I had just just downloaded and run the installer in response to a prompt to upgrade, so it's unlikely that my goal was solely to remove the old version. Furthermore, many installers give you the option of either removing the old version, replacing it with the new one, or just installing the new one so that you have a choice of which version to run. Since the remove button was highlighted as the default, that seemed to be the most likely interpretation.

Perhaps the designer of this dialog was aware of this confusion, and that's why it includes what Don Norman would call a "user manual": the wordy explanation that includes a sentence telling you what install and remove do.

What might have been done instead? First, we might note that the *install* concept has two distinct operational principles, one in which you install and use an app, and another in which you uninstall it and reclaim the space. If the intent were to support both of these, they might have been offered as separate workflows, perhaps in separate tabs (and with the installer shown as the default, especially when following a prompt to upgrade). Second, removing an

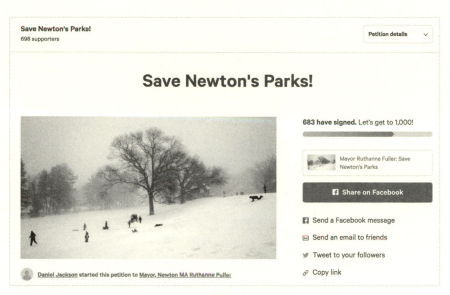

FIG. 8.2 *The petition owner's view of a change.org petition reveals a small deception: signers are shown a lower count (the 683 on the right in the body of the petition) than the actual count visible to the owner (the 698 in the admin bar at the top left).*

old version to replace it with a new one is very different from uninstalling. The former could be presented as an option ("remove old version?") next to the "install" button; the latter might be labeled "uninstall" rather than "remove."

Anyway, poor reader, I suspect you've heard enough of this unremarkable example. Suffice it to say that conceptual issues lurk, even behind a dialog box with only two buttons labeled with common English words.

Including a User Manual in the Interface

Sometimes a concept is sufficiently complicated that even the best designer will not be able to convey the meanings of actions or states without some additional explanation. In Chapter 2, I described how the Backblaze message "You are backed up as of: Today, 1:05 PM" was misleading due to the complexity of its *backup* concept.

It didn't mean that any file saved before 1:05 PM was safely backed up. To counter this interpretation, I might reword the message to "Last backup: Today, 1:05PM," and include a sentence underneath saying something like "This backup included all files saved prior to the scan that began at 12:48PM." Alternatively, and more conservatively, I might instead display the message "You are

FIG. 8.3 *An invitation to support a change.org petition. Or is it?*

backed up as of: Today, 12:48PM," and append below "This backup was completed at 1:05PM."

This is the approach Apple takes in the dialog for its *do not disturb* concept. Below the checkbox for "allow repeated calls," whose interpretation is far from obvious, a comment in a smaller gray font adds: "When enabled, a second call from the same person within three minutes will not be silenced."

Dark Patterns: Intentional Obfuscations

A company can nudge users to behave contrary to their interests, performing one action over another, or not performing an action at all. Often this is done with a mapping that intentionally (and maliciously) obfuscates the underlying concepts.[87]

The petition site, change.org, contains several such obfuscations. A few years ago, I created a petition to persuade our mayor not to locate a new city building in the middle of a local park. I noticed that every time I viewed my own petition (Figure 8.2), the number of signers seemed to be growing second by second, with the count rising as I watched.

Then I realized that, since I was the owner of the petition, a count in the top left of the screen, shown only to me, revealed the actual number of supporters. If I waited, after a few seconds the growing count would always shoulder off at that actual number. Every time the petition is displayed, the count starts below

FIG. 8.4 *Two buttons, one stacked above the other (and colored yellow and blue on the site itself), for apparently distinct actions of the Amazon Prime concept, but actually both bound to the same action.*

the actual number, and then immediately starts rising, giving a false impression of real-time activity.

More insidiously, after you sign a petition, you get a request to make a donation (Figure 8.3). Most people assume, quite reasonably, that this *donation* concept is synchronized with the *petition* concept in a particular way: that their donation is being passed to the petition organizer to help them fund their cause. In fact, the money pays for advertising on change.org (whose domain extension is misleading too: it is *not* a nonprofit). In the case of my own petition, supporters contributed over $2,000 to change.org. Had I understood the operational principle of "chipping in," I would have warned my signers against it.

Sometimes, a mapping obfuscates which action will occur when a button is pressed. On Amazon's UK website, I was offered a chance to sign up for a free trial of Amazon Prime, and presented with what seemed like two buttons: a yellow button labeled "Try Prime FREE" and a blue button below it whose label began with the word "Continue" (Figure 8.4), suggesting it might mean "continue without Prime." In fact, the full button label reads: "Continue with FREE One-Day Delivery." That is, *both* buttons activate the *signup* action! The option to continue *without* signing up is provided by clicking on the much less prominent blue link to the left of the button.

A mapping may control your behavior simply by withholding important information or making it hard to access. Many airlines, for example, make it hard to find the expiry date on your frequent flyer miles, hoping that they will expire

Primary	Social
☐ ☆ Alyssa, me 3	hacking meetups javascript

FIG. 8.5 *Labeling in Gmail: a conversation with two labels, "hacking" and "meetups."*

before you realize. (This is just one respect in which the *frequent flyer* concept often seems to be designed counter to the interests of flyers.) Likewise, PayPal has been accused of hiding users' account balances, which—combined with the lack of a synchronization that automatically transfers received funds to an external bank account—maximizes the balances that users maintain at their expense (and PayPal's profit).

Mapping Complex Compositions: The Mysteries of Gmail Labels

Google's email service, Gmail, provides the concept of *label* for organizing messages. For example, you might define a label *hacking* and affix it to messages in which you discuss programming with your geek friends. Then if you want to find an earlier message that you remembered was about programming, you could filter on that label.

This is a nice example of synergistic composition that I mentioned before (in Chapter 6). By using special "system labels" for sent and deleted messages, the *label* concept unifies all kinds of lookups. The button labeled "Sent" that takes you to a view of sent messages, for example, simply invokes a label query on the *sent* label.

Gmail also provides the concept of *conversation*. Its purpose is to group together the messages associated with a particular thread of discussion, so that you can see a message, its reply, the reply to the reply, and so on, together.

Composing and mapping these concepts together is challenging. The designers of Gmail chose to *affix* labels only to messages, but to *show* labels on conversations. This leads to some strange anomalies. In Figure 8.5, a conversation appears to carry two labels, *hacking* and *meetups*. And sure enough, if you filter on either label separately, that conversation appears. But if you filter on *both* labels, the conversation does not show (Figure 8.6).

Although surprising, this is not a bug. The labels shown against a conversation are the accumulated labels of all the messages it contains. In this case, one message in the conversation has the label *hacking* and another has the label

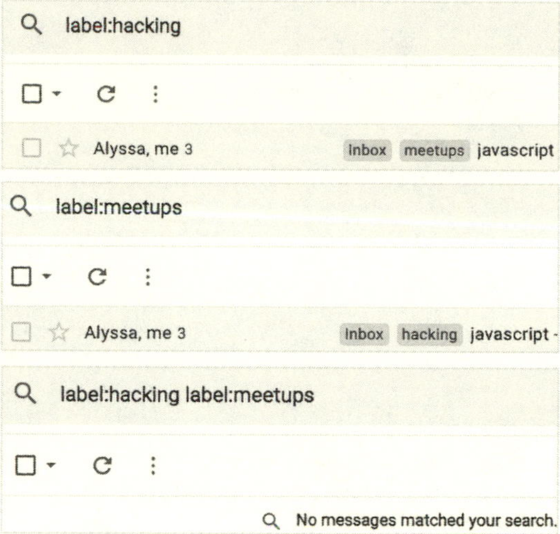

FIG. 8.6 *A label filtering surprise in Gmail: a conversation seems to match both the "hacking" and "meetups" labels, but is not shown if you query for both of them.*

meetups. So the conversation is shown as having both labels. But since no *single* message carries both labels, filtering on the two together yields no results.

You may wonder how different messages in a conversation could end up with different labels. When you select a conversation and add a label, it gets added to every message in the conversation. But you can define rules that affix labels to incoming messages based on their contents. And when you add a label to a conversation, it only affects the messages currently belonging to the conversation; messages added later do not automatically inherit the label. Moreover, some labels (such as the *sent* label) are applied automatically to individual messages.

In practice, the most troubling consequence of this design is more mundane: when you filter on a label, you get all the conversations containing messages with that label, but you can't tell which particular messages within a conversation actually carry the label.[88]

When you click on *sent* in Gmail to see the messages you've sent, for example, you get a list of *conversations* in which the sent messages are embedded, which includes messages that were not sent by you. The designers of Gmail mitigated this problem by showing the sent messages expanded by default and the rest compacted (Figure 8.7). But this distinction is not always easy to see.[88]

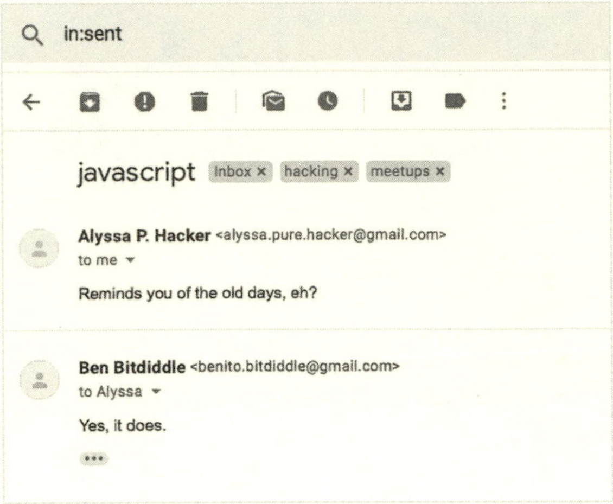

FIG. 8.7 *Filtering on sent messages in Gmail: the sent message is the first of the two.*

Worse, this default-expansion strategy appears to be applied only to sent messages. In other cases, all messages in a conversation are compacted except for the most recent message. This discrepancy is evidence that Gmail's designers are aware of this problem but have not yet solved it.[89]

Understandable but Unusable: Backblaze Restore

In all the examples we've seen so far, the problem has been a lack of clarity in the user interface: an uncertainty about the meaning of its controls and views in terms of the underlying concepts. Sometimes, in contrast, the meaning is clear enough, but the mapping makes it hard for users to perform the actions or get the information they need.

Backblaze (as I mentioned in Chapter 2) is an excellent backup utility that I've used for several years. The backups indeed run blazingly fast (up to about 200 GB per day), the setup is straightforward, and the service seems to be reliable. Moreover, restoring the latest version of a file is easy: you just go to the restore page of their website, select the file, and click to download it.

Restoring older versions, however, is not so easy. The dialog (Figure 8.8) lets you navigate the file system (on the left) to find the folder you're interested in, and (on the right) lets you select which files in that folder to download.

To restore an earlier version of a file, you can enter a date at the top of the dialog. There are in fact two dates. By setting the *from* date, you can include only

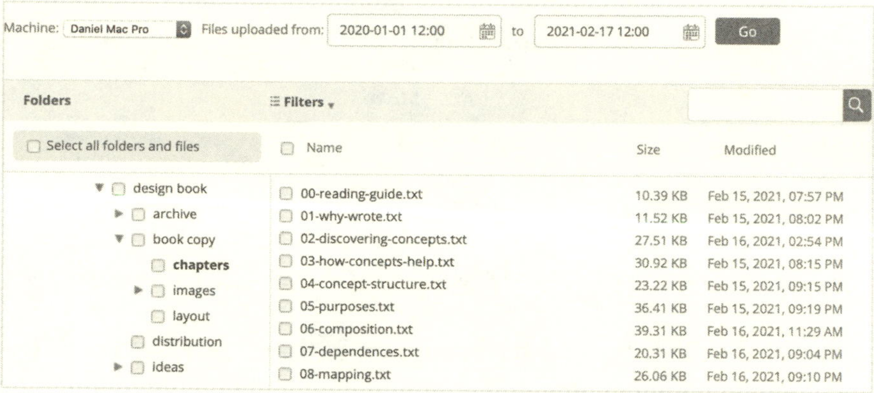
Machine: Daniel Mac Pro
Files uploaded from: 2020-01-01 12:00
to 2021-02-17 12:00
Go
Folders
Filters
Select all folders and files
Name
Size
Modified
design book
archive
book copy
chapters
images
layout
distribution
ideas
00-reading-guide.txt 10.39 KB Feb 15, 2021, 07:57 PM
01-why-wrote.txt 11.52 KB Feb 15, 2021, 08:02 PM
02-discovering-concepts.txt 27.51 KB Feb 16, 2021, 02:54 PM
03-how-concepts-help.txt 30.92 KB Feb 15, 2021, 08:15 PM
04-concept-structure.txt 23.22 KB Feb 15, 2021, 09:15 PM
05-purposes.txt 36.41 KB Feb 15, 2021, 09:19 PM
06-composition.txt 39.31 KB Feb 16, 2021, 11:29 AM
07-dependences.txt 20.31 KB Feb 16, 2021, 09:04 PM
08-mapping.txt 26.06 KB Feb 16, 2021, 09:10 PM

FIG. 8.8 *The Backblaze dialog for restoring files: the from and to dates filter which file versions are shown, a seemingly plausible mapping.*

files that were modified after that date; by setting the *to* date, you choose which version will be restored. For example, if you choose files *from* January 1, 2021 *to* March 1, 2021, you will be shown only files that were modified (or created) after the start of the year, and the version that you restore will be the last version that was backed up at or before the start of March.

This sounds fine, but there's a snag. Suppose you discover that an important file got corrupted somehow. You'd like to restore the last uncorrupted version. If you know on what date the corruption happened, you can just enter that date in the *to* box of the dialog. But if you don't know the date, you now have to search through old versions to find it.

You might start with yesterday, and move back one day at a time, restoring the file, and checking it until you find an uncorrupted version. If you know the corruption happened sometime between January 1 and March 1, you'd start with March 1, repeatedly restoring and checking until, in the worst case, you check 60 times and go all the way back to January 1.[90]

This might not seem like a big deal, but unfortunately, every time you change the *to* date in the dialog, you have to reload the folder tree (which takes about 20 seconds), and then, because the tree has been reset, you have to navigate down again to the file of interest. Then you have to download it and examine it.

Needless to say, this is a laborious process. The underlying concept is fine: the old file versions are all accessible; the problem is that the mapping makes it difficult to get your hands on them. One possible solution to this (adopted by other backup utilities such as Carbonite and Crashplan) is to show all the

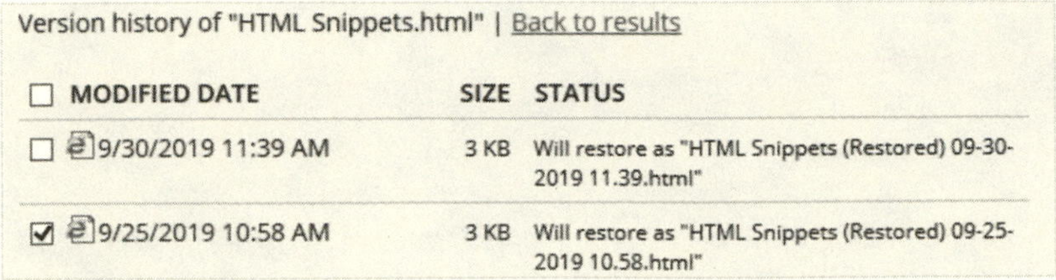

Version history of "HTML Snippets.html" | Back to results

MODIFIED DATE	SIZE	STATUS
☐ 📄 9/30/2019 11:39 AM	3 KB	Will restore as "HTML Snippets (Restored) 09-30-2019 11.39.html"
☑ 📄 9/25/2019 10:58 AM	3 KB	Will restore as "HTML Snippets (Restored) 09-25-2019 10.58.html"

FIG. 8.9 *The Carbonite dialog for restoring files, which shows a list of all versions of a file.*

versions of a file together with their modification dates, and let you download them in one go (Figure 8.9).

A Live Filtering Conundrum

Suppose an app allows you to display a collection of items, defined by some property that they all satisfy; and that, furthermore, while viewing the collection, you can modify the items. The design question is this: what happens if your modification of an item invalidates the property that made that item a member of the collection?

Apple Mail's *flag* concept fits this pattern: there are flags of seven different colors, which users can interpret as they please, and actions to attach a flag to a message, and to display all messages with a given flag.[91]

The *flag* concept maintains, as its state, a mapping from messages to the flags they carry. When you click on a flag icon in the sidebar on the left, the app displays all the messages that carry the flag of that color (Figure 8.10).

You might think, at first, that clicking on a flag maps to an action in the *flag* concept that finds all items with that flag. But it's better to think of the click as part of a more elaborate mapping that switches the interface into a mode in which the flagged items are displayed. This is better because it allows the display to be live, updating on the fly if the set of flagged messages changes spontaneously (for example, because a message just arrived and got flagged by a rule).

Now the conundrum. What happens if, while viewing the list of flagged messages, you select one of them, and remove its flag? A seemingly obvious solution is to remove that message immediately from the list, in order to ensure consistency (namely that the displayed messages are exactly those that carry the given flag).

FIG. 8.10 *A skillful mapping of the flag concept in Apple Mail: messages with yellow flags are being shown. The flag has been removed from the first message, but the message is still listed.*

In practice though, this would be a bad mapping design. Just think what would happen if you toggled a message accidentally, removing its flag and then wanting to reinstate it. When you removed the flag, the message would disappear from the list, your selection now cleared. You may not even be able to find that message to reinstate its flag—ironic, since you likely flagged certain messages *precisely* because finding them was hard![92]

A preferable behavior is, counterintuitively perhaps, to not update the display, and to retain all those messages that were originally shown. When you unflag a message, the message remains; but if you switch out of the flag view, and return to it later, then you will notice that the message no longer appears.

For this scheme to work, each message in the list must be flagged individually. This might seem gratuitous at first, since every message in the initial display must, by definition, be flagged. But when you toggle the flag on a message, you can see the flag disappear, and the message remains right there—and still selected—allowing you to easily reinstate the flag. This is exactly how Apple Mail behaves: in the screenshot, I have unflagged the message at the top, but it is still showing.

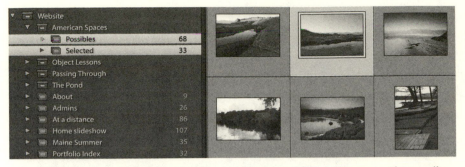

FIG. 8.11 *A mapping dilemma for the collection concept in Adobe Lightroom: with two collections showing, selecting a photo for removal does not identify which collection to remove it from.*

Resolving Ambiguous Actions

Often the user's gestures are easy to interpret. But sometimes they're ambiguous. This can happen especially when an action's arguments depend on previously made selections. Let's look at an example of this.

In the *collection* concept, items can be added and removed from collections that represent a possibly overlapping classification of items. Examples of applications that use collections are: Zotero, which lets you organize citations of papers into collections; browsers such as Safari, which offer collections for bookmarks; and Adobe Lightroom, which lets you define collections of photographs or movies.

The distinguishing feature of the *collection* concept that makes it distinct from the *folder* concept is that an item can belong to more than one collection. Mapping the action *collection.add(i,c)*, in which item *i* is added to collection *c* is usually straightforward, and may be accomplished (for example) by letting the user drag the item to the collection.

The action *collection.remove(i,c)*, in which item *i* is removed from collection *c*, is trickier. The problem is that, in some applications, more than one collection may be selected at once. This is an important feature, because it lets the user see the items belonging to multiple collections in a single view. In Figure 8.11, two overlapping collections of photos have been selected in Adobe Lightroom.

Indicating the arguments for the *remove* action is now no longer straightforward. You might expect to be able to select an item, and press the delete button (for example). But if the selected item belongs to both collections, it's not clear which to remove it from.

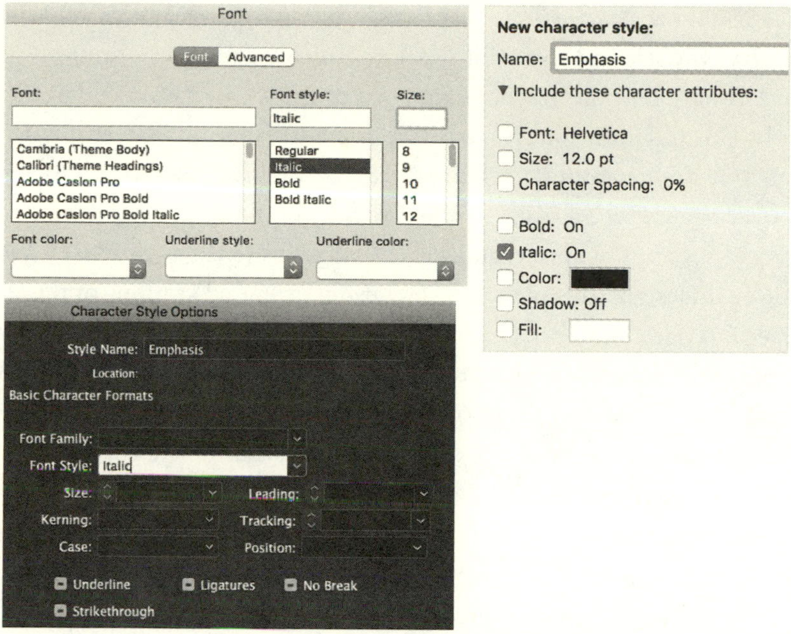

FIG. 8.12 *Mapping the "none" value in partial style dialogs. Recent versions of Word and InDesign (lower left) use extended widgets, such as dropdown menus whose selected entry can then be edited as text; an earlier version of Word had a separate selector for the font that populated an editable text field (top left); Apple Pages 09 (right) used checkboxes.*

This is a nasty mapping problem. When I first drafted this chapter (in late 2020), Lightroom displayed an error message if you tried to do this, informing you that the deletion request was ambiguous. Now (in February 2021), it simply deletes the item from *all* the collections that it belongs to that are currently selected.[93]

When Standard Widgets Aren't Enough: Entering No Value

Some concept actions take an argument that can either be drawn from some set of values, or can be "none," indicating that no value has been chosen. In the *format* concept, for example, you may have an action *set (p, v)* that sets format property *p* to have the value *v*; if you set a property and then wanted to *undo* the setting, you might then set it to the value *none*.

This subtlety arises in the partial styles example that I mentioned in Chapter 4. To recall, the idea is that you can define a *style* that specifies the values of only some formatting properties. For example, you might define a character style called *emphasis* that sets the font style to *italic*, but leaves every other font property (such as the size and choice of typeface) unchanged.

Now consider how this might be mapped to a user interface. We'll need not only to be able to set the font style to *italic*, say, but also to *unset* it: that is, to set the font-style value to *none*, its default. That's not the same as setting it to *roman*, which would cause type already in italics to be modified.

In recent versions of apps that offer partial styles (such as Microsoft Word and Adobe InDesign), this is achieved by using extended versions of the standard user interface elements (Figure 8.12, lower left). The boolean checkbox becomes a three-state widget (on, off, and not set), and the dropdowns are extended so that, in addition to picking a value from the dropdown menu, you can also *edit* the chosen value (as a text field) and thus delete it (to unset it). This is less counterintuitive than it might seem, since the text field can be used for autocomplete too, which is helpful when the menu contains many entries.

In earlier versions of these programs, the extended user interface elements were not available, and users had to tolerate clunkier interfaces or apply workarounds. In Word, for many years, a partial style could be unset only by writing a Visual Basic script; then at a later point, a complicated dialog let you choose a font from a list, which then populated a separate text field that you could edit (Figure 8.12, upper left). In InDesign, once you set a property, there was no way to unset it, except by using the "reset to base" action which would clear *all* the properties of the style. The iWork '09 version of Apple Pages (Figure 8.12, right) sidestepped the problem by adding a checkbox to each setting; this solution, while clean and clear, and not requiring fancy interface elements, was dropped, presumably because it required so much space.

All our previous examples in this chapter would apply in any user interface framework. This last example is interesting because it exposes a limitation of contemporary user interface toolkits, which generally do not offer a way to "unset" a value that was previously selected.

Lessons & Practices

Some lessons from this chapter:

· Concepts have to be *mapped* to a concrete user interface, with actions mapped to gestures such as clicking on buttons and concept state mapped to display views of various sorts.

- User interface design principles apply, but concepts help focus the mapping concerns. In the Java example, we saw how conflating installing and uninstalling in the same dialog causes confusion, and a mapping that paid more attention to the underlying concept structure would have been clearer.
- Some concepts are inherently more complicated, and call for more ingenuity in the mapping, and sometimes even explicit explanations in the user interface.
- Attempts to make the user interface simpler than the underlying concepts may backfire. In the Gmail example, labels are attached to messages, but the interface associates them with conversations, simplifying the visual appearance but compromising usability.
- The mapping must take into account typical usage patterns, which as the Backblaze and Apple Mail examples suggest, may be more complicated than performing a single action.
- Despite their expressiveness, visual interfaces may not resolve all ambiguities, and interface toolkits may constrain the mapping design.

And some practices you can apply now:

- When approaching user interface design, think first how each concept can be mapped individually. Then you can expand your view to consider how the mapped concepts fit together on the screen, and what transitions and links are needed between them.
- As you design a mapping for a concept, start by ensuring that each action is available in the interface (when relevant), and that the state of the concept is displayed (when needed) in an intelligible way.
- Check your interface design against the concept's operational principle and the most likely ways in which users will want to use the available actions.
- As always, a key advantage of a concept-based approach is that concepts, by factoring out generic functionality, make it easier to identify predecessors that can inform your work. So when designing the mapping for a concept, look to see how that concept is mapped in other apps.

PART III
PRINCIPLES

9
Concept Specificity

This chapter explains a simple principle that is surprisingly far-reaching in its ability to expose problems in a design. In fact, it is so simple that you may be tempted to dismiss it, but I hope the examples will convince you of its value.

The rule says that, in the design of a software product, concepts and purposes should be in one-to-one correspondence. That is, for every concept there should be exactly one purpose that motivates it, and for every purpose of the product, there should be exactly one concept that fulfills it.

You are probably not surprised that a concept should have at least one purpose: without one, what's the point? Or that a purpose that has been identified as significant for a product should have a concept that delivers it. That each purpose should be delivered, avoiding redundancy, by at most one concept is surely plausible too: why waste effort?

More radical is the suggestion that a concept should fulfill at *most* one purpose. And yet that aspect of the specificity principle turns out to be one of the most useful insights of concept design, and will take most of our attention in this chapter.

Concepts Without Purposes

A concept without a purpose is a strange beast, but we saw examples earlier (in Chapter 5): mostly cases, such as the editor buffer, in which an internal mechanism is exposed to the user. This section itself thus has no purpose, since I've discussed the topic at some length already.

Purposes Without Concepts

A critique of a design might reveal an essential purpose that has no corresponding concept to fulfill it. Since all software products evolve over time, new needs will always arise—but this is not what this category of design flaw entails. Rather, it points to concepts whose absence is egregious and obvious from the start.

If a concept is obviously missing, why would a designer not immediately add it? One reason is that it presents a challenge that is not easily resolved. For example, most email clients lack a concept of *correspondent* that would serve the purpose of identifying senders and receivers of email messages. This is easy to do within the confines of a closed email system such as Gmail, but to provide it more widely would require a universal authentication infrastructure.

With this concept in place, the sender field of an email message could not be forged, and spam would be much easier to control. It would also bring a more mundane but much needed benefit. In Apple Mail, you can't reliably search for messages from a given sender. The search bar misleadingly offers the option of searching by "people," but that turns out to mean nothing more than matching strings in the from- and to-fields of messages.

Most people have used more than one email address, so searching for them brings up multiple, distinct "people." And some format their name differently in their various email accounts, so you might need to search under different strings to find all the emails from a single person. In Figure 9.1, you can see that a search using my wife's name even brings up my own email address (probably because someone sent a message to my email address but with both our names in the to-field.)

Some more examples of purposes without concepts:

Deletion warnings in backup. Most backup utilities have a disquieting loophole in their terms of service (and in the behavior of their *backup* concept). Files that are deleted from the machine being backed up will also be deleted from the backup itself after some time period (say 30 days) has elapsed. The rationale for this policy is clear: it prevents customers from using the backup service as unlimited cloud storage.

It's ironic, however, that one of the main reasons people need backup is that they delete files by accident. It would be helpful and reassuring for backup utilities to provide a concept that tracks deletions, and warns you when they have occurred, so that you can determine whether or not they were intentional, before it's too late and the file has been cleared from the backup before you notice. The design of this concept is nontrivial, and would also have to account for renaming.

Missing style concept. Sometimes, a concept is routinely used in one class of application, and yet not available in another, even though it would be very

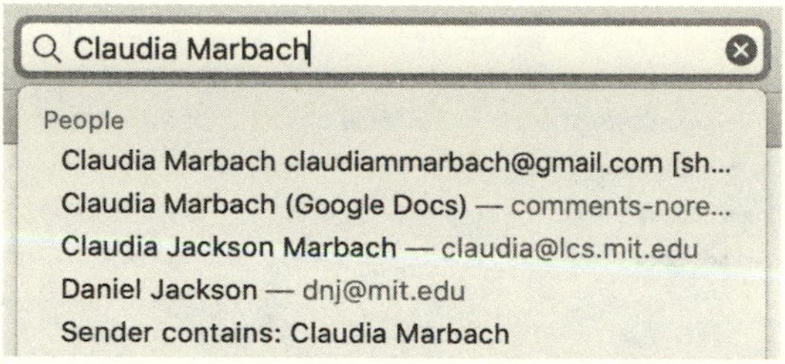

FIG. 9.1 *How lack of a correspondent concept in email leads to arbitrary search results.*

useful there. The concept of *style* is ubiquitous in word processors and desktop publishing tools, but was only recently introduced in Apple's slide presentation app, Keynote; it is still missing from Microsoft PowerPoint. Without styles, it is not easy to maintain consistent formatting, especially for texts such as formulas, code and quotations that are usually distinguished typographically.

Vestigial template concept. Occasionally, a concept is included in a design, but in such a limited form that it fails to fulfill its usual purpose. Most website-building apps include the concept of a *template* (sometimes called a *theme*), whose purpose is to decouple visual design from content. This lets you concentrate on the content of your site, and separately select a template that determines the layout, colors, fonts, etc. Key to this decoupling is that you don't need to commit to a template at the outset; you can start with any one that looks reasonable, put in some content, and then try others to see how your content looks with them.

This concept was implemented in Squarespace, but, inexplicably hobbled in version 7.1 released in early 2020, when they eliminated the ability to switch templates. This has left users in a bind because the new version brought several improvements (most notably a uniform data model across all templates, which one might have thought would have made template switching easier to implement).

With the proliferation of technology, it's easy to think that all real problems have been solved. These orphaned purposes suggest, reassuringly, that some of the most basic needs have yet to be fulfilled, and there is still important design work to be done even in the most familiar contexts.

📪 Primary		😃 Social 23 new	🏷 Promotions 100+ new	+
☐ ☆ ▷	**Google**	**New sign-in from Chrome on Mac** - New sign-in from Ch ✏	**12:30 pm**
☐ ☆ ▷	**Keith Muhammad at DeMont.**	**DeMontrond Auto Group** - 14101 North Freeway Housto	**12:19 pm**
☐ ☆ ▷	**AT&T High Speed Internet.**	**AT&T High Speed Internet Service Activation** - Your A1	**10:37 am**
☐ ☆ ▷	**Keith Muhammad at DeMont.**	**DeMontrond Auto Group** - 14101 North Freeway Housto	**Aug 26**
☐ ☆ ▷	**betterbatonrougejobs.com**	**Job Update -- 2015-08-26** - Looking For An Advantage W	**Aug 26**

FIG. 9.2 *Gmail's categories, a redundant concept.*

Redundant Concepts

A concept is redundant when another concept already exists to serve the same purpose. This may happen because the designer initially saw two distinct purposes, but they turned out to be variants of a single, more general purpose.

Gmail categories. Gmail introduced the concept of *category* whose purported purpose was to support the automatic classification of incoming emails (Figure 9.2). Rather than showing all the messages in a user's inbox in one list, the new concept offered a division into categories: "primary" (for email from personal contacts), "social" (for messages associated with social media accounts), "promotions" (for sales pitches), "updates" (for notifications, bills, receipts, etc.), and "forums" (for messages associated with groups and mailing lists).

You might have expected this new concept to have been greeted only with enthusiasm. After all, it provided powerful new filtering, promising to declutter inboxes and giving users more control. The new concept was met, however, with a barrage of negative articles and blog posts. And for a long time, a Google search for "Gmail categories" listed, as the first question above the search results, "how do I get rid of categories?"

The reason for this negativity, I believe, is that the *category* concept is redundant. Several of the critical blog posts noted that Gmail's *label* concept already met the same purpose: classification of messages. Moreover, that concept already included "system labels" (such as *sent*) that are attached without the user's intervention and could easily accommodate a new classification algorithm.

Why then didn't Google simply implement the new categories as system labels? Had they done so, users would not have had to understand the new concept of *category* and the apparently arbitrary restrictions that distinguishes categories from labels—for example, that only categories can be assigned to tabs, or that only labels can be used to classify messages outside the inbox.[94]

Zoom broadcast. In the Zoom videoconferencing app, you can move the participants of a call into "breakout rooms." The host of a call can send a message to all the participants using the concept of *broadcast*.

Why is this concept needed? Zoom already has a concept of *chat* that allows participants to exchange text messages throughout a call. The chat menu allows a participant to select whether a message is sent to a particular participant or to "everyone." Confusingly, in a breakout room, "everyone" takes on a different meaning and refers only to those in the room; there is no option to send a message to participants in other rooms. Moreover, the host cannot send messages to participants once they have been moved into breakout rooms; nor can participants address participants in other rooms. So the *chat* concept does not allow the host to send a message to breakout participants, which is where the *broadcast* concept comes in.

A better design might do away with *broadcast* entirely, and extend the *chat* concept so that chat messages can be sent across breakout rooms. Like Gmail's categories and labels, chat and broadcast messages are curiously incomparable even though they fulfill the same purpose: broadcast messages flash across the screen but chat messages appear in a rolling message log; broadcasts can cross breakout rooms but messages cannot; messages persist in the message log but broadcasts disappear. Most frustratingly for users, broadcast messages only appear for a few seconds, any links they include cannot be clicked on, and their content cannot be copy-pasted.

Ideally, the *chat* concept would offer the features of both concepts. Within a breakout room, the chat menu would let messages be directed to everyone in a room or everyone in the entire session; it would support messages to all participants individually, even if in different rooms; and, in particular, the host would be able to send messages to everyone in breakout rooms despite belonging to none of them.[95]

Search and rules in Apple Mail. Apple Mail has a search box that allows you to enter various properties of a message (such as text occurring in the body, or the name of a sender or receiver) and then filters the displayed messages accordingly (Figure 9.3). There is also a dialog for creating a rule that filters incoming mail that allows you to select messages satisfying certain criteria (such as whether the to-field mentioned you by name).

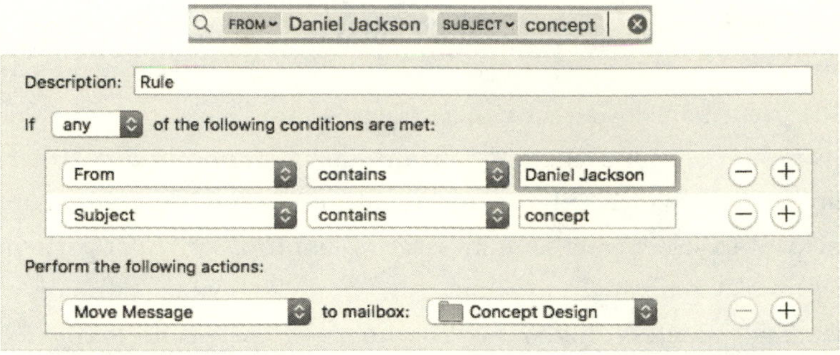

FIG. 9.3 *Apple Mail's filters (top) and rules (below), embody two versions of one concept.*

Underlying these two features is a common purpose, which we might call *message filtering*, that lets you define a subset of messages, in one case to be displayed, and in the other, to have some action applied (such as being moved to a folder). And yet this single common purpose is actually implemented twice within two distinct concepts, *rule* and *filter*, each time offering slightly different features. Rules but not filters can specify inexact matches (*contains* rather than *equals*, for example); rules but not filters can check the *cc* field; filters but not rules can specify that a message is signed; and so on. Gmail, in contrast, unifies rules and filters in a single concept.[96]

In all these cases, eliminating redundancy would have saved the developer work *and* given the user a simpler and more powerful tool.

Overloaded Concepts

Now the most interesting criterion: that a concept should have at *most* one purpose. A concept cannot serve two purposes well.[97] A purpose guides every aspect of a concept's design. If you have two different purposes, they will necessarily pull in different directions, and the concept design will have to compromise in favor of one over the other. More likely, the design will end up satisfying neither purpose in its entirety, because it is pulled in one way here and another way there.[98]

A concept that serves two purposes is *overloaded*. In the next few sections, I'll give examples of overloading, classified into four kinds by its cause:

· *False convergence* occurs when a concept is designed for two different functions that were assumed (wrongly) to be aspects of the same purpose.

- *Denied purposes* are ones that were ignored by the designer, despite the desires of users.
- *Emergent purposes* are new purposes for old concepts, often invented by the users themselves.
- *Piggybacking* occurs when an existing concept is adapted or extended to accommodate a new purpose.

Each of these overloadings has its remedy:

- False convergence is avoided by making an effort to articulate a single purpose precisely, and checking that different motivations for the concept are truly reflections of the same purpose.
- Denied purposes are avoided by taking seriously the opinions and experience of users, especially those who are less technical and more reluctant in their technology adoption.
- Emergent purposes are the hardest to avoid, since nobody can predict all the ways in which a design will impact the context of its use and create new uses. Just recognizing emergent purposes as they arise, though, allows them to be addressed, say by adding new concepts to accommodate the new purposes.
- Finally, piggybacking is avoided by developing an awareness of the urge to "optimize" a design by putting concepts to contradictory purposes, and learning that the effort saved in such reuse leads to a complexity that eventually exacts a high price.

Overloading by False Convergence

Sometimes two different purposes of a concept seem to be so well aligned that the designer treats them as a single, converged purpose, until it becomes clear that in fact the purposes are not only distinct but may be inconsistent with one another.[99]

In Facebook, for example, you might describe the purpose of the *friend* concept as "allowing two users to establish a relationship in which they can see each others' posts." The problem with this formulation is that it hides two distinct purposes. One is *filtering*: by promoting posts from your friends, Facebook saves you the trouble of sifting through posts from people you're not interested in. The other is *access control*: by choosing your friends, you can choose who gets to see your posts.

For most users these purposes are indeed often aligned because human relationships tend to be symmetrical: if I'm happy for you to see posts about my personal life, I'm probably interested in seeing posts about yours too. But for celebrities, the symmetry breaks down. I might want to read Barack Obama's posts, but I doubt he would want to read mine.

Recognizing this, in 2011, Facebook added the concept of *follower*, which serves only the filtering purpose and not the access control purpose. The *friend* concept still plays both roles, but you can use it solely for access control by turning off "following" for friends whose posts you don't want to see.

Overloading by Denied Purpose

Like false convergence, *denied purpose* involves a second purpose that existed at the time of the initial design. But in this case, the designer rejected the purpose, deciding that it did not merit recognition in the design.

Listing and then rejecting candidate purposes is often admirable. It's a key strategy for preventing bloat in application design. It's so tempting to build a Swiss Army Knife application that solves every possible problem, but the result is not likely to be a happy one. The agile mantra to build The Simplest Thing That Works applies both in selecting purposes to address and in the design of the concepts that fulfill them. Sometimes, though, omitting a purpose can be an act of obstinacy and denial, ignoring the needs of the user in the name of purity of vision.

Twitter's *favorite* concept (discussed in Chapter 5) is an example of this. Prior to 2018 when Twitter introduced the *bookmark* concept, its users had no way to save a tweet except by making it a favorite, and revealing this choice publicly. So the *favorite* concept was forced to serve two incompatible purposes: signaling approval and saving tweets for later.

I have generally avoided talking about programming tools because they're not familiar to most people, but here is an example I hope I can explain convincingly. Programmers use version control systems such as Git, Subversion and Mercurial to manage team work on a codebase, and to keep track of multiple versions of files. In practice, though, many users—especially less expert ones—use these tools for backup too.

Just think about it. All these systems allow you to frequently copy your work from your machine, which might fail at any time, to a server, and to maintain

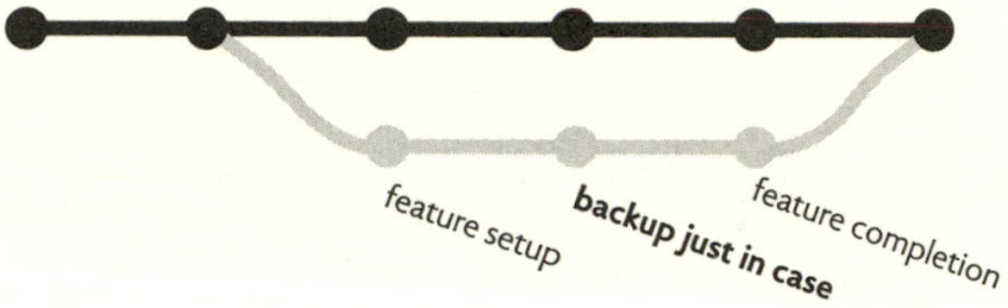

FIG. 9.4 *Example of denied purpose: using commits for backup. The lower, gray path is a separate "branch" in which a programmer builds a new feature. She commits her work when she has set up the basic structure of the feature, and finally when it's complete. But she also commits in the middle at an incoherent point because she wants to backup her unfinished work.*

multiple versions of each file going back in time, which you can revert to at any point. Isn't that exactly what a backup system does? If you're using such a system already, why wouldn't you use it to back up the same files that are already being stored in the repository?

Unfortunately, the designers of these tools generally do not share this view. The concept that is relevant here, the *commit*, was designed for a different purpose, namely for storing snapshots of a project that correspond to coherent states of development. For example, you might perform a commit when a feature is complete, or when your work is in a sufficiently polished state to benefit from peer review.

This purpose is not compatible with the backup purpose, because you want to back up your files as frequently as possible. If you do a large piece of unfinished work, you would certainly want to back it up, but it may not be in a coherent state. So you have a dilemma. If you commit the work product, you will be misrepresenting it, and (as developers say) "polluting the commit graph" by inserting commits that have no coherent meaning and may not even compile (Figure 9.4). But if you don't commit it, it won't get copied to the cloud, and you risk losing it if your machine fails.

Overloading by Emergent Purpose

A concept may have a single, compelling purpose at the time of its design, but may acquire additional purposes later as users discover new uses for the concept. The story of the plain old subject line of email messages is a case in point.

The *subject line* concept can be seen as an instance of a more general concept, *precis* say, whose purpose is to make it easier to find, filter and assimilate long

> To: csail-related@lists.csail.mit.edu
> Re: [csail-related] turn off the lights?

FIG. 9.5 *Subject line used for emergent purpose: identifying listserv origin.*

texts by giving a short summary that is created along with the original text or added later.

With these modest origins, one might not have predicted the many glorious roles the *subject line* came to play. Listserv software added a prefix to the subject line with the name of the listserv, making it easier for recipients to tell that the message was not sent directly to them, apparently replicating the purpose of the *to* field (Figure 9.5). Later, email systems such as Gmail started using the subject line as a heuristic for grouping messages into conversations.

These newly emerging purposes might seem harmless, but they're not. A friend told me of a case in which someone in his department sent a message to multiple colleagues. They blind-carbon-copied the recipients, including an email address of a listserv for a handful of departmental officers. The listserv duly revealed itself as a target of the message in the subject line, undermining the privacy guarantee that the *bcc* concept is supposed to provide.

The use of the subject line for conversation grouping is a known problem, because it spuriously associates messages that happen to have the same subject. One of my students told me that he likes to assign different labels to messages associated with different trips, so he can see all the messages associated with a given trip by filtering on the respective label.

Unfortunately, his travel company uses "your upcoming trip" as the subject line for all confirmations, causing messages about different trips to belong to the same conversation. And since filtering on a label in Gmail brings up all messages in the same conversation as a labeled message—a design flaw I explained in Chapter 8—he can't use labels to show the messages for one trip without seeing messages for other trips too.

Overloading by Piggybacking

The most common reason for overloading is that a designer sees an opportunity to use an existing concept to support a new purpose, avoiding the need for a new concept—and thus saving the trouble of designing and implementing it. The designer might also imagine that users will appreciate the economy of

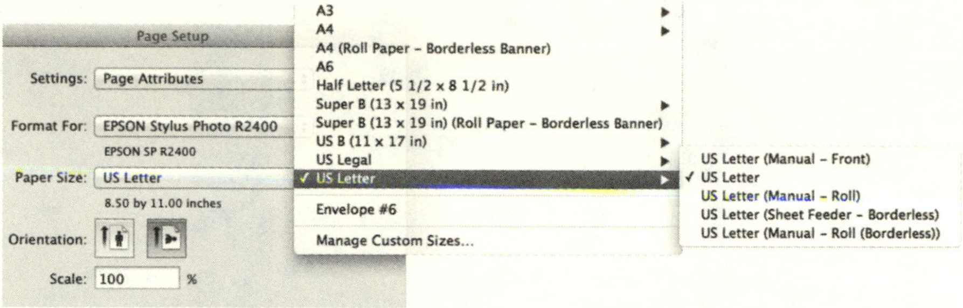

FIG. 9.6 *Epson printer driver: paper feed piggybacked onto paper size.*

fewer (and richer) concepts, but this is often illusory. Better to have more concepts that are coherent and compelling than fewer concepts that are complicated and confusing.

Epson's unusual concept of paper size. In Apple's macOS operating system, a printer driver can offer printer-specific settings in the print dialog. Epson photo printers offer many categories of special settings, allowing you, for example, to choose the paper type, adjust the drying time between prints, and so on.

These printers typically offer several different feed options for different kinds of paper: from the top, from the back or the front, from a roll of paper, and so on. How is this *paper feed* concept controlled?

Apple has a built-in concept of *paper size* that is set in most applications through the page setup menu. The purpose of *paper size* is to make it easy to define standard paper sizes just once, and then reuse them. In addition to the built-in paper sizes (such as standard letter size), you can define custom sizes of arbitrary dimensions, along with margins. When you choose a paper size in an application, it sizes the page appropriately (wrapping text and respecting margins). And when you print the page, the paper size is passed to the printer so that it can check that it matches the size of the paper that is loaded in the printer.

Sadly, rather than adding *paper feed* as a new concept, Epson choose to piggyback it on the *paper size* concept. When you open the page setup menu, the names of the paper size options include the paper feed setting (Figure 9.6)! This might seem to be a small hack but it creates major havoc:

· When you choose the paper size for a document that you intend later to print on your Epson printer, you have to commit to one of these specialized

FIG. 9.7 *Shooting with a square aspect ratio: a great feature of mirrorless digital cameras.*

options. You would rather not commit even to a particular printer (so that you can later choose which printer to use in the print dialog), let alone a feed option.

· Because the feed option is hard-wired into some standard paper sizes that come with the printer driver, you can't create new custom paper sizes, because the feed options are not available as a user-customizable setting.

· Some applications use the page setup for defining presets. Adobe Lightroom has a concept of *printer preset* that lets you define borders and layout for a given paper size. You might, for example, define a postcard preset for creating a postcard from a photo. Because the preset relies on the paper size, if you want to print to an Epson printer, you need to choose one of the piggybacked paper sizes that includes the feed. As a result, your printer presets are now feed-specific.[100]

Fujifilm's aspect ratios. Fujifilm cameras let you set the *aspect ratio* of the image in camera. The purpose of this concept is to allow you to frame the image in the viewfinder at shooting time with a particular aspect ratio for the final image. You can choose to shoot square images (Figure 9.7), for example.

To set the aspect ratio, you open the ominously named *image size* menu (Figure 9.8, left). You can see that I've chosen a square (1 × 1) ratio, but, strangely, I also had to choose the image resolution—in this case, "L" for large.

A different menu called *image quality* (Figure 9.8, middle) lets you choose how the photo is recorded on the memory card: as just a raw file, or as a JPEG (in normal or fine quality), or as a combination.

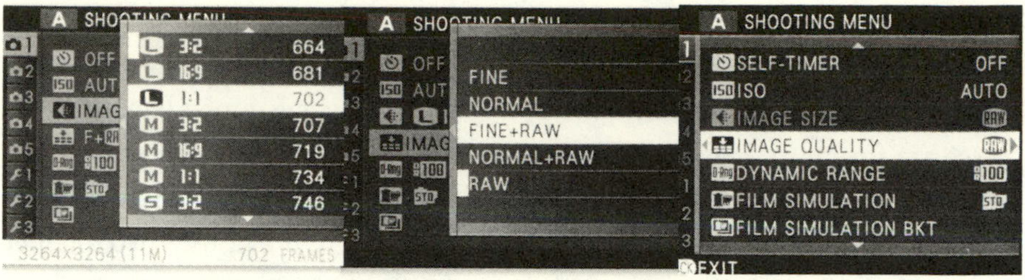

FIG. 9.8 *Piggybacking in Fujifilm cameras: Setting image size/aspect ratio (left); setting image quality (middle); choosing "RAW" as quality greys out the size option, and with it the ability to set a custom ratio (right).*

Suppose you want to take square photos and save them only as raw files. If you switch the image quality to raw, you'll find that the image size setting is grayed out and replaced by the word "RAW" (Figure 9.8, right). This makes sense for the image sizes—such as the large setting that we chose—because they don't apply to a raw file, which always includes all the sensor pixels. But why is our aspect ratio lost now too?

There's no good reason. The aspect ratio concept works perfectly well on raw files (elegantly saving the crop outline as editable metadata), but because it's tied to JPEG dimensions through overloading, it can't be applied to them alone. In practice, this means that if you want to just record your images in raw, and you want a custom aspect ratio, you have to choose the image quality setting that includes JPEG files with the raw files, and then delete the JPEG files later!

The remedy is simply to provide an *aspect ratio* concept that is distinct from the *image size* concept. Making these orthogonal would make custom aspect ratios independent of the file type. The menus would also be simpler: instead of the nine entries needed for all the combinations of three image sizes and three aspect ratios, there would be two menus of three entries each.

This might seem like a small detail, but I wonder if it's behind Fujifilm's reluctance to offer more aspect ratios, which many users have requested (there's even an online petition). Without a distinct *aspect ratio* concept, however, the quadratic growth in menu options would make the image size menu grow unacceptably as new ratios are added.[101]

Purpose Granularity & Coherence

Whether a design exhibits redundancy or overloading will depend on how purposes are formulated. You may be wondering whether this isn't a rather subjective and arbitrary judgment. What if we just attempted to resolve overloading by declaring a new purpose that merges what were previously two distinct ones? Would that eliminate the problem of having two purposes for a single concept? Of course not: we need some test of *coherence* to reveal when multiple purposes are masquerading as one.

Ideally, a purpose will be formulated without separating cases, so that its coherence is evident in the wording itself. For example, when I talked earlier in this chapter about the *template* concept used by website-building apps, I said the purpose was to "decouple visual design from content." This formulation implicitly unifies a variety of activities without suggesting multiple purposes.

But suppose I had instead described the purpose as "making it easier to build a website with attractive visual design by letting you start with a template designed by someone else, and then changing the template later without having to start again." This is certainly an inelegant purpose, and by referring to the details of particular actions it's beginning to resemble an operational principle. But it's not wrong to present this as a single purpose, because the various parts are aspects of a single and coherent purpose (albeit better summarized in terms of decoupling). Indeed, this was my criticism of the Squarespace design, which treats each part as if it were a purpose in itself.

If a purpose is expressed in multiple parts, then, how are we to know whether it is coherent? Here are some criteria:
- *Reformulation*. Is there a compelling reformulation without multiple parts?
- *Common stakeholders*. Do the benefits of each part accrue to the same stakeholders?
- *Common mission*. If we identify a higher-level purpose for each part (which we might call a "mission" to distinguish it from the concept's immediate purpose), would the parts have the same mission?
- *Non-conflict*. Are the parts non-conflicting, or can we imagine a scenario in which a user might want one but not the other?

Let's take a look at an example to see how these criteria can be applied.

FIG. 9.9 *The like concept in Facebook: the emoticons have tooltips that identify them as "like," "love," "care," "haha," "wow," "sad," and "angry."*

Applying Coherence Criteria: Facebook Likes Has Multiple Purposes

When you click on the "like" button below a Facebook post, seven emoticons are displayed, offering different sentiments from love to anger (Figure 9.9). If we were to treat this a concept—let's call it *like*—what would the purpose of the concept be?

Several things come to mind. Most obviously, clicking on one of these emoticons conveys an emotional reaction from you back to the author of the post (albeit publicly). This is probably what most Facebook users have in mind as they click away.

But there's more going on. If you play around with Facebook, you'll discover that what you like affects which posts you're shown, and the order in which they appear. By indicating which posts you like, you are curating your feed for the future, making it more likely to contain posts you will want to see.

Less helpfully, your clicks are tracked by Facebook to build up a profile of personal data that is used to target you with advertisements. From the content of your posts, and your reactions to the posts of others, Facebook classifies you on a host of metrics from hobbies to sexual orientation.

In summary, we might say that the purpose of the *like* concept is to "convey emotional reactions, curate your newsfeed, and provide tracking data for targeted ads." Regarding these as three distinct parts, we can now apply our criteria.

Reformulation. "Reacting to a post" sounds plausible, but it is not need-focused, and thus not a purpose (as explained in Chapter 5). I'm not sure I can do better though.

Common stakeholders. Facebook would no doubt argue that selling your data to advertisers to allow them to target you more effectively benefits you as well, by showing you more relevant advertisements. But most of us would resist this claim, and see the advertisers and Facebook itself as the beneficiaries. In contrast, curating your newsfeed is for you alone; and conveying emotions is a benefit shared with a larger community. In short, it seems that each part benefits a different group of stakeholders.

Common mission. Likewise, the missions of the parts diverge. Emotional reactions serve the building of relationships and communities; curating your newsfeed serves your need for more engaging and informative content; and tracking for advertisers serves Facebook's bottom line.

Non-conflict. Finally, the parts are in conflict too. Certainly most users would like to be able to curate their feed, but would prefer not to be tracked. Sending reactions is not well aligned with curating your feed either; you might want to send a supportive gesture to a friend but prefer not to see more of their posts. The "angry" emoticon is especially confusing: are you being angry *with* the author of the post (expressing support for their outrage), or are you angry *at* the post? It seems that this reaction is used in both ways, but according to Facebook, all of the reactions have the same effect in terms of curating your feed, and indicate (as the main button suggests) that you "like" the post, even if it makes you "angry."

In summary, our compound purpose is revealed by the coherence criteria to be multiple purposes, and the Facebook *like* concept is therefore an example of overloading.

Splitting a Concept: Facebook Likes Should Be Multiple Concepts

The remedy for overloading is to split a concept up, with one new concept for each purpose. In this case, we could split *like* into three concepts: *reaction*, whose purpose is to covey emotional reactions to posts; *recommendation*, whose purpose is to allow you to curate your feed; and *profiling*, which is used to construct an advertising profile for targeting ads.

When making such a split, the most encouraging sign is that the new concepts already exist. And, indeed, the *reaction* concept (reactions without curation) appears in communication apps such as Slack and Signal; the *recommendation* concept (curation without reactions) appears in Netflix, where a thumbs-up or thumbs-down influences which movies are suggested; and the *profiling* concept is used by Google's Gmail service to target advertisements based on the content of your email messages.[102]

By treating these as three distinct concepts, we can now explore varying degrees of synchronization. At one end of the spectrum, we might have a free composition with almost no synchronization at all. In this version, users would have to click separate buttons for each of the concepts. This is not entirely implausible, because it would give users complete control, but it would not suit Facebook's interests, since very few people would click on the profiling buttons.

At the other end of the spectrum, the concepts would be fully synchronized, so that clicking on an emotional reaction corresponds also to upvoting the post for curation and contributing to your profiling. This of course is how Facebook is designed now. The problem of overloading has been converted into a problem of over-synchronization. But at least with the concepts separated there is a clearer indication in the design that the concepts have been coupled together, and less risk that the concepts themselves will be corrupted in an attempt to navigate between conflicting purposes.

Between the two extremes, there are other design points. One option would be to continue to hide the *profiling* actions, but to tease apart *reaction* and *recommendation*, with one collection of buttons for sending a reaction, and separate buttons for upvoting or downvoting a post.

In fact, Facebook users have asked for a "dislike" button, and Facebook has turned the suggestion down, arguing that dislikes would introduce a negative spirit into the platform. This seems disingenuous, and it assumes that the *like* concept has not been split. With a split into *recommendation* and *reaction*, you could dislike a post (executing the *recommendation.thumbs-down* action) *without* sending any social signals.

No doubt Facebook's designers have considered all these factors and more. What concepts bring is a new framework in which to analyze a design and make principled trade-offs. Splitting concepts is valuable in large part because it allows an idiosyncratic concept (such as Facebook's *like* concept) to be decomposed into concepts that are more coherent and more familiar—and thus a better basis for a comprehensible user experience, and a better structure for recording and preserving design knowledge.

Lessons & Practices

Some lessons from this chapter:

- The specificity principle says that concepts should be one-to-one with purposes. This simple rule has some profound implications for concept design.
- Concepts without purposes are rare, but can arise from exposing to the user mechanisms that should have been hidden.
- Purposes without concepts to fulfill them may indicate constraints that originate outside the designer's domain, or sometimes just egregious omissions.
- Redundancy, when multiple concepts serve the same purpose, confuses users and wastes resources.
- Overloading, in which a single concept has multiple purposes, arises in several ways: from false convergence, in which the designer mistakenly imagined that multiple purposes were actually one; from denied purposes, in which the designer intentionally ignored a purpose for which users then marshal an existing concept; from emergent purposes, in which concepts find new (and often incompatible) purposes over time; and from piggybacking, in which the designer attempts to save design and implementation effort by hitching a new purpose onto an old concept.
- Violation of any of these results in an increase in complexity and a loss of clarity: purposeless concepts needlessly clutter the interface and confuse users; missing concepts lead the user to work around the omission with more complicated interactions; redundant concepts introduce confusing distinctions between two concepts that should be one and the same, and force the user to learn different ways to do the same thing; and overloaded concepts bring unexpected complexity from the coupling of unrelated purposes.
- Unwelcome restrictions on functionality often result too. Redundancy is often a symptom of concepts appearing in specialized contexts, perhaps by subteams, without the kind of design attention that is paid to the core concepts of an application. And overloading leads to restrictions because the second purpose is forced into the procrustean bed of an existing concept.
- Coherence criteria help determine whether a purpose expressed as multiple parts is really one purpose or several. They include: whether reformulation as a single purpose is possible; whether the parts have common stakeholders;

whether they serve a common mission; and whether they conflict with one another.

And some practices you can apply now:

· When you're designing an app, ask yourself early on if there is an essential purpose that you have not even considered. In analyzing feedback from users, consider whether any problems that users have might be due to an omitted concept.

· Compare each of your concepts pairwise to make sure you have no redundancy, and look more deeply to see if there are shared functions of concepts that could be factored out into their own, common concepts.

· If a concept becomes complicated, or doesn't seem to work intuitively and flexibly for your users, it may be overloaded. Formulate a purpose as precisely as you can, using the purpose criteria of Chapter 5; and if you can only express the purpose in multiple parts, apply the coherence criteria of this chapter to determine whether the parts reflect distinct purposes.

· When you determine that a concept is overloaded, try to split it into more coherent concepts, each with a more compelling and unified purpose. Be on the lookout for standard concepts that you have seen elsewhere; a composition of familiar concepts is more flexible and powerful than a single, idiosyncratic concept.

10

Concept Familiarity

Design novices often imagine that expert designers have an uncanny ability to pull brand new ideas out of thin air. But what looks like momentary inspiration is more often insight drawn from years of experience. A great designer has a repertoire of designs in mind, ready to inform each new design problem she encounters. Only when the standard solutions prove inadequate will she reach for something new.[103]

In this respect, software is no different from any other design field. To apply the lessons of previous designs, you need first to be able to extract design ideas as reusable fragments. That's what concepts aim to do. A concept is a particular solution to a particular design problem—not a large and vague problem, but a small and well-defined need that arises repeatedly in many contexts.

Inventing a new concept to fulfill a purpose for which a perfectly good concept already exists isn't just a waste of effort. It also tends to confuse users who are likely to be familiar with the existing concept. In this chapter, we'll look at some examples of this. But first the good case—when familiar concepts are successfully reused.

Successful Reuse of Concepts

Concept reuse is rampant, especially in web apps. In fact, it sometimes seems as if every social media app is fundamentally the same—a variant of a ubiquitous super-app that lets you make connections with people and communities, share text, images, and video with them, and react to the contributions of other users with comments and ratings. Viewed from a little distance, many of the popular apps—Facebook, Twitter, Instagram, WhatsApp, SnapChat,, etc.—seem to be barely distinguishable, varying only in small details.

When yet another app in this category appears, you might initially be puzzled by how it differs from existing apps. But you won't have any trouble figuring out how to use it, because it will likely offer all the concepts you are already

FIG. 10.1 *The Twitter user with the most followers.*

familiar with: *post, message,* and *comment* for creating content; *friend, follower,* and *group* for accessing and filtering it; *rating, upvote* and *moderation* for quality control; *notification, favorite,* and *recent activity* for highlighting content; and so on.

The same concept may appear under different guises. So the old *chatroom* concept becomes a *group* in WhatsApp or Google Groups or Facebook, and a *channel* in IRC or Slack. Twitter provides a nice example of connecting its design to an existing concept. Here, for example, is how it explains the concept of *follower* (Figure 10.1):

> *What does it mean to follow someone on Twitter? ... When you follow someone, every time they post a new message, it will appear on your Twitter Home timeline.*

In answering the question, Twitter provides the concept's operational principle. It is exactly what a Twitter user needs to know. The meaning of following someone isn't explained in terms of some abstract notion of liking them or even wanting to read their tweets. It's given as a simple scenario: you follow, they post, you see the message on your timeline.

But I omitted a crucial part of Twitter's explanation (marked by the three dots). The full answer begins:

> *Following someone means you've chosen to subscribe to their Twitter updates. When you follow someone...*

The first sentence is what's relevant here. Saving you the trouble of learning the concept if you already know it, it notes that *follower* is just a form of the familiar *subscription* concept, in which you subscribe to some set of events—in this case tweets from a given user—and are notified when they occur.[104]

148

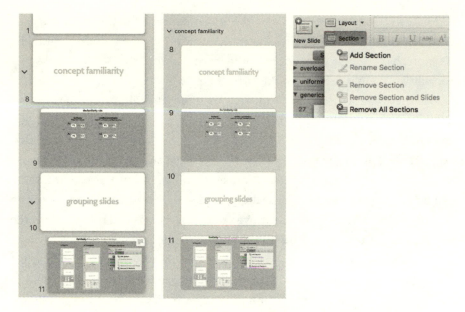

FIG. 10.2 *Organizing slides in Keynote and PowerPoint. On the left, the Keynote group concept, which reuses the familiar concept of the tree outline; in the middle, the PowerPoint section concept, which is novel and unfamiliar; on the right, some actions on sections whose behavior is unpredictable.*

Grouping Slides: Avoiding Invention

When a designer has a choice between reusing a generic concept or inventing a new one, it's preferable to reuse the generic one—unless there's a clear respect in which the generic concept would not fulfill the purpose so effectively.

To illustrate this, let's look at how two slide presentation tools let you structure the sequence of slides. The purpose here is to be able to organize a presentation into smaller groups of slides, so that you can work on each group separately. The expected operational principle is something like this:

If you group a contiguous set of slides, you can then apply actions to the entire group at once, such as showing or hiding the slides, moving them around, etc.

Apple Keynote provides the *slide group* concept for this purpose. A group is a sequence of slides that sits under a parent slide, shown by indenting the members of the group with respect to the parent (Figure 10.2, left). You can toggle the visibility of the group, and can move it around as a unit by dragging the parent slide.

Microsoft PowerPoint provides the *section* concept for the same purpose (Figure 10.2, middle and right). Each section can be given a name, and, like Keynote, has a toggle to display or hide its constituent slides. This works reasonably well, and is certainly not a terrible design.

But the Keynote design, in my view, is more effective and easier to use. Whereas PowerPoint's sections are limited to one level of organization, you can put groups inside groups (up to six levels deep). In these sample slides, you can see that, in Keynote, slide 11 sits below the slide with the title "grouping slides," which itself sits below "concept familiarity"; in PowerPoint, there can be no structure within the section entitled "concept familiarity" because sections can't be nested. You can move sections forwards and backwards in the presentation order, but you can't put a section inside another section.

The user interface of groups is more intuitive. To create a group, you select some slides that follow the slide that will be the parent, and you drag them to the right; to remove the group, you drag them back to the left. If you drag a slide that is in the middle of a group to the left, it will be promoted up one level (as you'd expect), and the group it previously belonged to will be split into two sibling groups.

Creating sections is more tricky. You might have thought you could select a contiguous sequence of slides (say all the slides that are about concept familiarity) and then invoke the command "add section." If you do that, it will indeed create a section, but that section will include all the slides from the first selected slide to the end of the entire presentation. If the selected slide is within a section, the new section will follow it; if not, a second new section (which will be named "default section") will be created for the earlier slides.[105]

This is complicated! (And tedious to read about, I admit.) But worse, it's unpredictable. There's no reason you should have been able to guess that adding a section would do this. It would have been equally reasonable (and perhaps better) to create a new section containing just the selected slides, for example.[106]

In contrast, Keynote's behavior is mostly simple and predictable. If you start with no groups, and you drag one slide to the right, you will have created *one* group in which that slide is the child and its predecessor is the parent. Unlike with sections, no other groups appear spontaneously. Perhaps the only behavior you might not predict is what happens if you select several non-adjacent slides and drag them to the right. In that case, you see visually (as you begin to drag) that the slides are grouped together into a single contiguous sequence, and will then become children of the same parent.

Why was Apple able to design a better concept? In large part, it's because they didn't start from scratch. And this is also why Apple's concept feels more intuitive. We've seen it before, in other contexts. We might call it the *outline tree* concept. Every outlining tool and word processor has it: you can make a list of items (usually short sentences or phrases) and introduce levels, so that the resulting structure is a tree with an outline item at each node.

Export Presets: When Extension Breaks Familiarity

Our second example of an unfamiliar concept arises in a different context. In this case, we'll see that the design started out with a conventional and familiar concept, but it was then extended with new functionality, and the familiarity was lost.

The concept is *preset*. Its purpose is to save the user the trouble of entering parameters for a frequently used command. Instead, the parameters are saved as a preset, and when the command is invoked, the user can choose either to set the parameters explicitly, or to set them automatically by selecting a previously saved preset.

Adobe Lightroom Classic uses presets effectively for many different commands. There are presets for printing, for editing, and for importing and exporting images. The troublesome example that I want to focus on involves the use of the *preset* concept for exports.

Take a look at the screenshot of the export dialog (Figure 10.3). On the right are the parameter settings; on the left is a (hierarchical) list of presets. You can adjust the parameters manually; you can also click on the name of a preset, which will cause the parameters to take the preset values, and you can then override those values if you like. All this is familiar to anyone who has used a preset dialog.

But now if you look carefully at the preset list, you'll notice that there is checkbox next to each preset name. This, it turns out, is a powerful extension that allows you to select several presets at a time. Now you may wonder: what can this possibly mean? Well, in general, for the *preset* concept, it doesn't make much sense, because only one set of parameters can be used when a command is executed. But for this particular command, one might want multiple

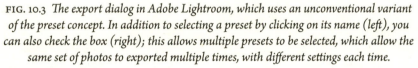

FIG. 10.3 *The export dialog in Adobe Lightroom, which uses an unconventional variant of the preset concept. In addition to selecting a preset by clicking on its name (left), you can also check the box (right); this allows multiple presets to be selected, which allow the same set of photos to exported multiple times, with different settings each time.*

executions with different presets. You might, for example, export the chosen photos in both high and low resolution versions in one go.

The goal of this new functionality—namely to allow a sequence of exports with multiple presets—is totally reasonable, and was apparently requested by many users. But squeezing it into the *preset* concept produces some strange anomalies. For example, if you select a preset by checking it, you can no longer edit its parameters.

As you might expect, when this new functionality was added, many users posted requests for help on the app's community forum. Many (including me) didn't even realize that clicking on a preset name was different from clicking on the checkbox. And the graying out and the hiding of sections was puzzling to users. The presence of the "Learn more" link at the bottom of the dialog suggests that Lightroom's designers are well aware of these problems, but have yet to solve them.

Applying the concept specificity principle, we might detect here two distinct purposes: (1) saving common parameter settings for commands; and (2) repeating a command with different, but predetermined, settings. The first is fulfilled by the *preset* concept. The second probably calls for a new concept, which would be independent of *preset* but used with it: maybe one like Photoshop's *action* to define sequences of actions as little programs that the user can define and invoke.[107]

Conformity of Concept Instances

When a concept that appears in a design is an instance of a familiar, generic concept, it should adhere exactly to the behavior of the generic concept—unless there are very good reasons not to, and the deviations from the standard concept are made very clear. Otherwise users who are familiar with the generic concept will be confused, having assumed that the concept behaves the same way it behaves in the other instances they have seen.

To illustrate this, let's look at a dilemma in the design of Apple's *contact* concept. Most people use the Apple Contacts app on their phone. Aside from storing phone numbers so you don't have to remember them, it also fulfills the useful function of attaching names to numbers when calls come in. Many people enter nicknames for friends and family members, so that on a certain Charles George's phone, a call from his mother might be shown with the name "Mummy" if he had failed to follow protocol and enter her name in his contacts as Her Majesty Queen Elizabeth II.

There's no disaster here, but if the Prince of Wales now sent an email message to his mother, he might be distressed to discover that the email would include "Mummy" in the recipient email address of the message. Awkwardly, names attached to email addresses are passed on when a user forwards or replies, so if the message were about some matter of state that involved advisors, it might make its way around all the offices of Buckingham Palace until eventually every message sent to the Queen is addressed to "Mummy."

If the Prince were to make this mistake, I would argue that we would have to forgive him, on the grounds that he legitimately assumed that the Contacts app used a concept that we might call *nickname* that lets you use a convenient alias or nickname in place of a longer phone number or email address. That concept keeps the alias private, so from that perspective the behavior of the Contacts app is deviating from familiar expectations.

In Apple's defense, we might argue that the concept never was *nickname*; from the start it was a *contact* concept that lets you store all the information you have about someone, including their full name. The Prince was misled by the fact that he happened to look up contacts by name, but Apple's app does lookups and autocompletion equally on phone numbers and email addresses. His expectations were also shaped by using contacts for phone calls (in which no name is ever sent) before he used contacts for emails.[108]

In this case, there is no right or wrong answer; the point is simply that familiarity with a concept, and the concomitant expectations it brings, are powerful factors that must be taken seriously in design.

Lessons & Practices

Some lessons from this chapter:

- A good designer knows not only how to invent new concepts, but also when not to invent at all. If your purpose is addressed by an existing invention, you're better off reusing that.
- A concept is like any other invention in this respect. What's new is that concepts provide a way to structure knowledge and experience of software design in small and coherent pieces, allowing more granular opportunities for reuse.
- The easiest way to make a design usable is to build it from familiar, preexisting concepts. Using polished and well-understood concepts reduces the chance of misfits, and makes the design intuitive to users.

And some practices you can apply now:

- Before you invent a new concept, brainstorm existing concepts to see if there is one that meets your purpose. Remember that the concept you need may come from a very different domain.
- When mapping to a user interface, a need for unconventional widgets may suggest that the underlying concept is itself baroque and unconventional.
- If an existing concept seems to only partially meet your goal, rather than modifying or extending it, explore whether it might be composed with another familiar concept to give the functionality you need.

· When the behavior of concept actions is unpredictable, and several possibilities seem equally likely, the concept design is likely at fault. A good design has a quality of inevitability about it.

11

Concept Integrity

When a system comprised of concepts executes, each concept runs as its own little machine, controlling when an action may occur, and what its effect on the concept state will be. Synchronizations can constrain actions further, by making the actions of one concept happen together with certain actions of another concept.

One concept cannot modify the state of another concept directly, or somehow change the behavior of one of its actions. This is critical, and what makes concepts intelligible in their own right.

But this modularity only holds if concepts are properly composed, using the synchronization mechanism of Chapter 6. If the framework in which the concepts are implemented allows them to interact in other ways, or if there are bugs in the code, a concept may behave in an unexpected way, violating its specification.

The designer can also break a concept, tweaking its behavior so that, in composition with other concepts, it conforms to the needs of the particular app. Some adjustments might preserve a concept's specification while adding some new functionality, but others might break it.

For all these reasons, it is critical that the *integrity* of a concept be maintained when it is composed with other concepts. In this chapter, I'll show you some examples of integrity violations and the problems they cause.

Some integrity violations (such as our first one, The Revengeful Restaurateur) are blatant and easy to fix once discovered. Some (such as the second, Font Formats) are subtle and represent an ongoing design struggle that has yet to be resolved. Some (such as the third, Google Drive) are unsubtle but fixable only with considerable effort.

A Blatant Violation: The Revengeful Restaurateur

Imagine a restaurant reservation app with a *reservation* concept with actions to *reserve* and *cancel* tables, and a *review* concept that lets users post ratings of restaurants they've visited.

Each of the two concepts has its defined behavior and its operational principle: for *reservation*, that if you reserve and turn up at the right time, a table will be available; for *review*, that aggregate ratings reflect the individual ratings that were previously submitted.

When these concepts are composed, the designer can synchronize them together. For example, she might decide that you can't review a restaurant until you've reserved it (or maybe even dined there). This synchronization will constrain the app by ruling out certain behaviors—in particular, ones in which a user reviews a restaurant they never made a reservation for. Despite the synchronization, every behavior of the app will still make sense when viewed through the lens of a particular concept.

Now suppose a restaurateur, frustrated by bad ratings, decides to hack the app to punish ungrateful customers. He modifies the behavior so that a customer who enters a bad rating is able to make a subsequent reservation, but then finds—even though there was never a *cancel* action—that when they arrive at the restaurant there is no record of the reservation, and thus no table.

This hack does not correspond to any legitimate synchronization. Not only does it couple together the two concepts, but it *breaks* the *reservation* concept. The operational principle of that concept says that if you make a reservation and don't cancel it, a table will be available. With this hack, the principle no longer applies, and the app cannot be understood in terms of the original concept. This is what I will call an integrity violation.

Suppose, on the other hand, the revengeful restaurateur hacked the app so that when any customer posts a low rating, a *cancel* action is performed on any reservation the customer has at any restaurant. The poor customer probably gets a notification (due to synchronization with the *notification* concept) of the cancellation, despite never intending to cancel.

This behavior, however mean-spirited and annoying it might be, does *not* violate integrity because the new behavior is perfectly understandable in terms of the specification of the *reservation* concept. It might annoy the customer to

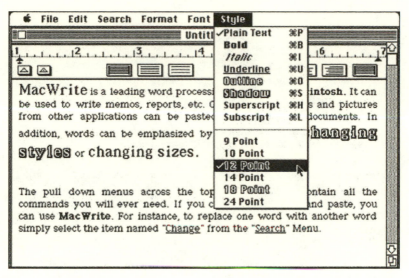

FIG. 11.1 *The format toggle concept in the first versions of MacWrite (1984).*

discover that a cancellation has been issued without their consent, but the behavior is still consistent with the concept (its specification being silent on the question of *who* is allowed to cancel a reservation).

Font Formats: A Long-Standing Design Problem

In the first word processors, text was formatted with three simple properties: bold, italic, and underline (Figure 11.1). Each property had an associated action that toggled it, so if you applied the action *bold* to plain text, it would become bold; and if you applied it again, it would return to plain. This concept is so familiar and remains so widely deployed that it seems silly to have to name it. But for the sake of our discussion, let's call it *format toggle*. You can find it today in thousands of apps from email clients to embedded rich text editors.

Another important (and early) concept for formatting text is *typeface*. Its behavior is simpler: there's a list of typefaces, and you can choose one and apply it to some text. In the early days, the *format toggle* concept was implemented as a transformation that was applied to the characters provided by the *typeface* concept: a character was italicized by applying a slant to the letter form, and made bold by a different transformation that increased the weight.

Real typographic italics, however, have never been just slanted versions of the roman forms, but are typically more flowing and calligraphic; nor are the

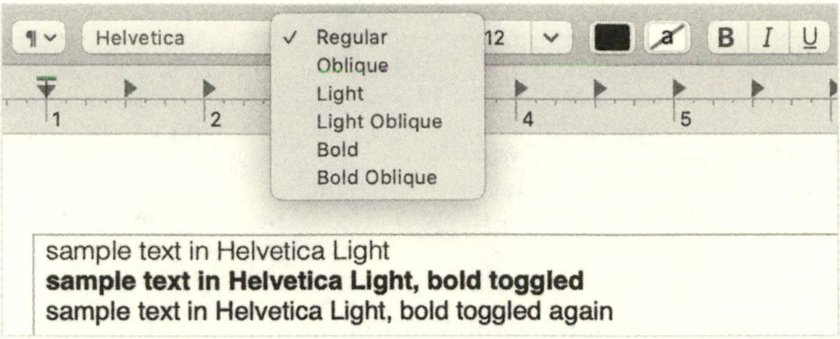

FIG. 11.2 *Integrity violation example in TextEdit: bolding once (second line) turns the text from light to bold; bolding again (third line) leaves the text in regular, not light.*

bolder versions of type just fatter. Computer typography advanced, and with the advent of PostScript fonts, it became common to provide distinct bold and italic versions of the typeface in separate font files, and to use transformations only for scaling. The implementers of word processors were able to maintain both concepts, *format toggle* and *typeface*, by a clever trick. When you set some text to italic, it switched to the italic font file; setting it to bold would then switch to the bold-italic font file; setting it to italic again would then switch to the bold font file; and so on. In this way, the design preserved the integrity of both concepts.

Then, with the arrival of professional fonts, trouble hit. Now, instead of just having a few variants of each typeface, a much larger collection was provided. The difference between these and the old fonts is usually additional weights such as semibold (between roman and bold) and black (heavier than bold), as well as additional variants for use at different sizes, such as a display font (for text set in very large sizes), or a caption font (for text set in very small sizes).

With these enrichments, all hell breaks loose, and *format toggle* no longer works. Figure 11.2 shows what happens in Apple's TextEdit. You can see I've selected the typeface family Helvetica, which has six variants. The first line was set in the Light variant. I then copied the text to the second and third lines. To the second line, I applied the bold action once, and to the third line I applied it twice. If *format toggle* works correctly, applying the bold action twice should take you back to where you started, so the first and third lines should look identical. But they don't, because applying bold once changed the type from Helvetica Light to Helvetica Bold, and applying it again changed it to Helvetica Regular (and not back to Helvetica Light).

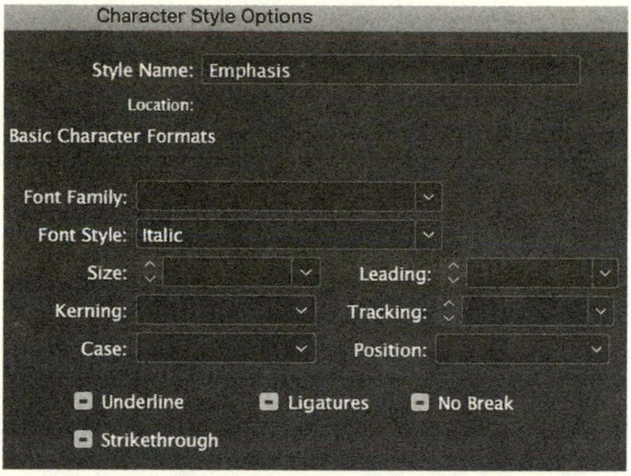

FIG. 11.3 *The character style dialog in Adobe InDesign: formats are specified by select-ing styles such as Italic and Bold, which undermines the value of partial styles.*

In short, the implementation of *format toggle* in TextEdit does not meet its specification, but not because there is a bug in the code. The problem is a deep-er one, and involves the interaction between the two concepts. The extension to the *typeface* concept has broken the *format toggle* concept.

Apple tried to fix this problem in its productivity apps such as Pages. The dialogs looks just like TextEdit, but the bold and italic actions behave differ-ently. If you bold some text in Helvetica Light, it will now be in Helvetica Bold (naturally); if you bold it again, however, it will be back in Helvetica Light (in accordance with the specification of *format toggle*). But this behavior is achieved with some hidden magic, which introduces new problems.[111]

This critique might seem nitpicky, but it's actually a serious problem in desk-top publishing. Figure 11.3 shows the character style dialog in Adobe InDesign. Here, I'm defining a style called *Emphasis* to be used for text that is to be empha-sized. By making it a style, I am hoping to be able to factor out *whether* some text is emphasized from *how* it is emphasized (by italics, bold or even underlin-ing, say). For an initial definition of the character style, I've selected the "font style" *Italic*. Note there is no selection for the "font family"; this is essential, because it allows the character style to be applied to text in different typeface families.

At least that was my hope; in fact, it doesn't work. To apply this *Italic* setting, InDesign switches the typeface to the one whose name is the typeface family concatenated with the string "Italic." So if the text is in "Times Regular" it will set it to "Times Italic." So far so good. But if the text is in "Helvetica Regular" it will try to set it to "Helvetica Italic." As you can see from the TextEdit screenshot (Figure 11.2), my version of Helvetica calls the italicized form "Helvetica Oblique." So the character style is *not* in fact typeface-independent, and can only be applied successfully to text in certain typefaces.

There have been other attempts to fix this problem, but there seems to be no satisfactory solution. The *format toggle* concept just cannot be reconciled with more sophisticated typographic concepts.

Losing Your Life's Work with Google Drive

My wife keeps most of her work documents in Google Drive. Having seen accidents in Dropbox (Chapter 2), I was worried about her losing her work, and began looking for ways to protect it.

I learned that Google Drive itself does not provide backup,[112] so I would have to devise my own scheme. An obvious idea came to mind. I would install the Google Drive app and keep all of her cloud files synchronized to a folder in her local disk, and would then add that folder to the selection set of the backup utility that I already had running on her laptop. That way, whenever one of her Google Drive files was modified, the local version would be updated, and would then be backed up to the cloud.

I was surprised to discover that this apparently straightforward scheme does *not* work. Searching online to see if anyone had come up with a solution to this dilemma, I came across a sad story of someone who had relied on a variant of this scheme and paid a heavy price.

The story is illustrated in Figure 11.4. On the left is the starting state, in which there are two files, *book.gdoc* (a Google document) and *book.pdf* (a PDF export of the document), both stored in the Google cloud and synchronized to the Google folder on the local disk. Our protagonist then moves the files out of the folder on the local disk, resulting in the state shown in the middle. The Google Drive synchronizer then runs, and seeking to make the contents of the local folder and the cloud folder identical, it removes both files from the cloud.

At this point, you might imagine that, whatever happens to Google Drive, the files are safely stored on the local disk. Sadly, this was not the case. As our hapless user reports:

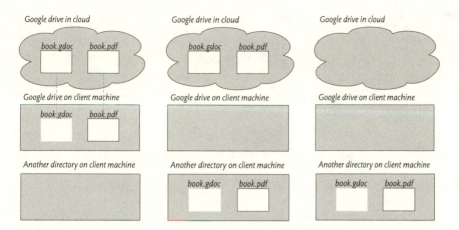

FIG. 11.4 *Integrity violation in Google Drive: the cloud-app concept breaks the synchronization concept. A user moved files out of his Google drive in order to make space in the cloud, but the files he moved turned out just to be links to files in the cloud that no longer existed.*

> *The next morning, I go to open a .gdoc file and get this error: "Sorry the file you have requested does not exist." My heart sank. What happened to the work from yesterday? I opened another file. Then another. All of them the same message. I was starting to freak out.*

Indeed, most of his files were gone, for good.[113] His summary: "I lost years of work and personal memories that I saved as Google Docs files because of a poor user interface." As we shall see, though, the problem was deeper than the user interface: it was a concept integrity violation.

Our user was relying on the behavior of *synchronization*. The purpose of this concept is to maintain consistency between two collections of items; the operational principle is that any change made to one collection is propagated to the other. Synchronization, unlike backup, also propagates deletions; this allows you to keep items organized. A fundamental property of synchronization is that the copies of the items in the two places should be identical.

Unfortunately, the Google Drive synchronizer does *not* always create faithful copies. It does for conventional files, such as *book.pdf*. But for Google app files, such as *book.gdoc*, it doesn't copy the file's data to disk at all. Instead, it creates a file that contains just a link to the file in the cloud. That's why attempting to open the file on the local disk produced an error message: clicking on it opened a web page in the browser for a file in the cloud that no longer existed.

In addition to *synchronization* then, there's another concept at play, which we might call *cloud app*. This concept embodies the idea of documents in the cloud that are accessed through a link. In concept terms, combining the two concepts has violated the integrity of the *synchronization* concept.

From a concept design point of view, there is no obvious barrier to fixing this problem (in contrast to the case of the *format toggle* concept). I suspect it's just not a priority for Google to implement a solution, although it's surprising that more users of Google Apps aren't more concerned about not having backups.

Lessons & Practices

Some lessons from this chapter:

· When concepts are composed to form an application, they may be synchronized (as explained in Chapter 6) so that their behaviors are coordinated. This synchronization may eliminate certain behaviors of a concept, but can never add *new* behaviors inconsistent with the concept specification.

· But if the concepts of an application are assembled incorrectly, behaviors may result which, viewed in terms of the actions and structure of a particular concept, break that concept's specification.

· These *integrity violations* confuse users, because their mental models of concept behavior are broken.

And some practices you can apply now:

· When designing an app using concepts, even if you are not defining synchronizations precisely, at least convince yourself that every interaction between concepts can at least in principle be viewed as a synchronization.

· If you're having trouble using an app, or analyzing a usability problem, and you discover that a concept is behaving in an unexpected way, ask yourself whether interference from another concept may be to blame.

· To ensure integrity, make sure that a concept that purports to be generic really is. In the Google *synchronization* example, the integrity violation is evident in the non-uniform way in which different types of files are handled.

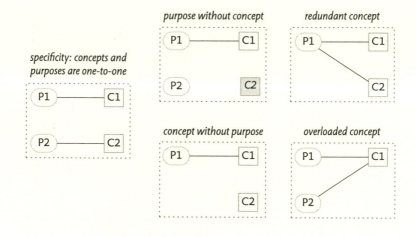

specificity: concepts and
purposes are one-to-one

purpose without concept

redundant concept

concept without purpose

overloaded concept

familiarity: when the same purpose
arises in different apps, the same
concept is used to fulfill it

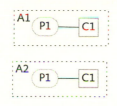

familiarity violation: different
concepts are used for the same
purpose in different apps

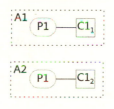

integrity: when composed,
each concept still fulfills its purpose

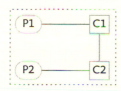

integrity violation:
one concept breaks another

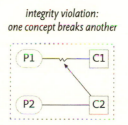

FIG. 11.5 *A pictographic summary of the principles of Chapters 9 to 11. A line between a purpose and a concept indicates that the concept fulfills the purpose; the broken line (for the integrity violation) indicates non-fulfillment, due to the interference of another concept; lines between concepts denote composition; dotted boxes represent applications.*

12

Questions to Remember

In closing, I'd like to review the key ideas of the book, and suggest how readers in different roles might apply them. The suggestions are organized around a series of questions.

For Strategists, Analysts, and Consultants

For those strategizing about a product and its evolution, the identification of concepts and their value dominates, with the design details of individual concepts taking a back seat.

What are the key concepts?

Consider the system, service, or application to be built—or the one that already exists—and ask yourself what its key concepts are. By constructing an inventory of the concepts, you'll get a bird's-eye view of the functionality, a kind of landscape within which to consider your strategic moves. Arrange the concepts in a dependence diagram to see how they are related to one another, and which ones lie at the core.

How old are your concepts?

When you look at the concepts in your existing system, determine when each concept was introduced, and investigate whether it has changed over time or remained stable. Are there concepts that have morphed dramatically (like Facebook's *post*—see Note 48) signaling a major shift in the entire system or becoming new concepts as they evolved? Have some concepts been introduced and then retired? Which have most successfully stood the test of time?

What are your most valuable concepts?

Do you have a killer concept (like Photoshop's *layer* or the World Wide Web's *url*) that is responsible for the success of your product and your competitive advantage? Are some concepts (like Gmail's *label*) the linchpin of your product without which it could barely function? Are some concepts key to revenue, perhaps because they define a premium version of your product, or because they bring the greatest value to your customers?

Do you have troubled concepts?

Does your product include concepts that have confused users, as evidenced by frequent help requests, or whose complexity has led to a disproportionate share of defects or system outages? If so, are these troubled concepts shared with your competitors, or might they be self-inflicted wounds?

What shared concepts define this family of products?

If you view several of your products as members of a single family (such as Adobe Creative Suite, or Microsoft Office), can you identify the key concepts that are shared between them? Are these shared concepts implemented using a common infrastructure, or are they implemented anew in each product? Are the various instances of a shared concept consistent with one another, or are there small and perhaps arbitrary differences between them? Do those differences create problems as users move from one product to another? Do they cause integration and data sharing problems?

Maybe the products belonging to the family don't currently share concepts, but they could in the future if concepts appearing in multiple products were unified. Would such a unification bring benefits not only to the family as a whole, but also to the individual products?

What is the purpose of each concept?

For each concept in your inventory, can you give a simple and compelling purpose? Do these purposes contribute to the larger goals of the product, and the vision of your organization?

Whose purpose does each concept serve? Does the purpose serve the interests of your customers? If so, which customers does it serve—the users or the advertisers? If it serves the interests of your organization, does it exact an unnecessary cost from customers? Are the purposes that are intended to serve the interests of customers communicated effectively to them, and do they align with their true needs?

Are there missing concepts?

Can you identify a purpose that is not being fulfilled that suggests a missing concept (such as the *correspondent* concept missing from email clients)? If you can identify such a concept, is there an opportunity to add it to your product, and thus acquire a competitive advantage?

What are your competitors' concepts?

Look at the competing products in the same domain, and inventory their key concepts. Do they differ from yours? Are the concepts that you have and that your competitors lack significant? Do they give your product an advantage, or are they a source of needless complexity? Are the concepts that your competitors have but that you lack a threat to the future of your product? Have you adopted industry-wide concepts? If so, do they make it easier for new customers to start using your product? Or do these concepts trap you in the flawed assumptions of past products?

For Interaction Designers & Product Managers

Many of the questions that apply to strategists and consultants also apply to interaction designers and product managers, but there are new questions focused on the design and mapping of individual concepts, and on tracing usability problems back to concepts.

Are the concepts consistently conveyed to users?

Does your product succeed in projecting—through its interface, user manual or help pages, training and marketing materials—a mental model that matches the actual concept model? Review the way that your product's functionality is described in the user interface and all supporting materials. Do these all present a consistent image of the product's concepts? Is there a common vocabulary for concepts and their purposes?

How are concepts explained?

Are the product and its associated support materials organized systematically around concepts? In your support materials, do you explain what the purpose of each concept is? Do you sometimes fall into the trap of explaining in detail what a concept does, without explaining what it's *for*? Do you provide compelling usage scenarios, and do they highlight the operational principles that show convincingly how the design of each concept fulfills its purpose?

What kind of usability problems do you have?

Reviewing feedback from users and requests for technical support, can you identify the primary usability problems in your product? Then, for each problem, can you determine what *kind* of problem it is, assigning it to one or more of the three levels of interaction design?

Which concepts make you happy or sad?

As a designer, you undoubtedly have a deep understanding yourself of the product and its qualities. Make a table for aspects of the design that are successful, problematic, or something in between. When you've filled up this table, review each item, assign it to a design level, and for all those items that turn out to be conceptual, name the concept that is responsible.

Are any of your concepts redundant?

Can you find redundant concepts (like Gmail's *category* concept) that serve the same purpose of other concepts in your product, giving you the possibility of simplifying and clarifying the design by eliminating a concept, and extending the other (if need be) to cover the functionality of the eliminated concept?

Are any of your concepts overloaded?

Do you have concepts (like the *cropping* concept in previous versions of Photoshop—see Note 101) that seem to serve multiple purposes? If so, these might be the cause of usability problems. Can you find scenarios in which the different purposes of a concept conflict with each other? If not, can you formulate a coherent and compelling purpose that encompasses those apparently distinct purposes, thereby arguing that the concept is, in fact, not overloaded after all?

Can some of your concepts be split?

Look at your more complicated concepts, especially those that are overloaded, and consider whether you might split them (as we did with Facebook's *like* concept) into multiple concepts, each with a simpler and more compelling purpose. Might doing this give you an opportunity to use a concept more widely and uniformly in your product? For example, if you factored a *notification* concept out, could you provide notifications of a wider class of events, and give the user control over which notifications occur?

Are familiar concepts used effectively?

For each of the concepts in your product, ask yourself whether there is a more familiar concept that might take its place. Are any of your concepts (like the *section* concept of Microsoft PowerPoint) close in purpose to existing, more familiar concepts? If so, would anything be lost by replacing them with their more familiar counterparts? And if you determine that your use of an unfamiliar concept is justified, are the ways in which it diverges from more familiar concepts made clear and understandable to users?

How are concepts composed?

Which concepts are tied together by synchronizations? Can you draw synchronization diagrams to show which actions are tied together? What kinds of compositions do your synchronizations achieve: free, collaborative or synergistic? How much of the power of your design comes from synchronization, and how much from the concepts themselves?

Do you have under-synchronizations?

Are there cases in which you could spare the user some manual work by increasing the synchronization between concepts, so that some actions are performed automatically? Could such synchronizations be provided as defaults for naive users, and in a more customizable form for experts?

Do you have over-synchronizations?

Are there cases in which concepts are synchronized too tightly, taking too much control from the user? Would more orthogonality between concepts (that is, a looser synchronization) give the user finer control, making available functionality already present in your concepts?

Are you exploiting synergy?

Do your existing concept compositions create synergies (as with the *trash/folder* example), in which one concept amplifies the power of another? Can you find additional opportunities for synergy? One way to think about this: could you adjust the behavior of one concept, perhaps generalizing it slightly, so that it could incorporate some of the behavior of another concept, but offer that behavior more consistently and widely?

Are the concepts mapped effectively to the user interface?

Does your user interface present concepts transparently to the user, or are the concepts buried under a layer of complex controls that makes it hard to see them and keep them separate? Is it easy for the user to discover how to select actions and their arguments? Is the state of each concept visible to the user? Does your user interface make not only individual concept actions available, but also more complicated sequences of actions that users are likely to need?

Have you analyzed your concept's dependencies?

Construct a dependence diagram for all the concepts in your product. Is the justification for each concept solid in terms of the concepts it depends on? Does the diagram suggest subsets you had not considered, perhaps for simplifications of the product?

Are the concepts assembled with integrity?

Each concept may be sound in isolation but undermined when combined with other concepts in the product as a whole. Does the design preserve the integrity of each concept? Or are there subtle ways in which a user's understanding of a concept has to be modified due to interference from another concept?

Is your concept wisdom safely documented?

A concept design may evolve over many years, accruing a host of fixes and refinements from multiple generations of designers. If this knowledge is captured only in the code, then—as the fate of the *range* concept in Apple Numbers suggests—it can be lost when a new programmer is unaware of subtleties, and makes a change that erases years of insight in seconds. For this reason, it's important to maintain a design journal for a product that tracks the development of each of its concepts. A briefer concept catalog or handbook that records the distilled wisdom of each concept the company has designed can enable sharing between products and help bring new designers up to speed.

For Technical Writers, Trainers & Marketers

Some additional questions apply to those who provide the crucial materials that users turn to in order to become familiar with a product and figure out what to do if they get stuck.

Are the supporting materials organized around concepts?

Are the user manuals, help functions and technical support articles organized around the key concepts? Are the actions of a concept explained together, in a coherent way?

Do you give clear purposes for concepts?

When introducing a concept, do you explain *why* the concept exists, what it's *for*? Do the purposes you give satisfy the criteria of well formed purposes (cogent, need-focused, specific, evaluable)? Have you avoided misleading metaphors?

Do you explain the operational principle of each concept?

To explain how to use a concept, do you give a compelling operational principle, or do you just list actions and leave the user to figure out what the archetypal usage scenario is?

Are concepts explained in a rational order?

If some of your materials (a user manual, for example) are sequential, do they present concepts in an order that is consistent with the dependence diagram, so that each concept can be motivated at the point at which it is introduced without forward references to concepts that have yet to be explained?

For Programmers and Architects

The above questions about the concepts, their purposes, and their relationships to one another are all fundamental for implementers too. The dependence diagram can be used to phase development incrementally, and to plan partial releases.

What set of concepts forms a minimum viable product?

This is of course a vital question for strategists too, but it has particular significance for implementers because they can more readily assess the costs of building the concepts.

Which concepts will be challenging to implement?

Can you identify which concepts will be the most challenging to implement? Which concepts have the most complex state, or will present performance challenges because of the volume of data they will embody? Do the operational principles of any concepts hint at consistency problems that may require distributed consensus algorithms? If so, might eventual consistency suffice?

Can you avoid reinventing the wheel?

If you're implementing a familiar concept, can you find implementations of that concept in your own organization or elsewhere that will give you guidance and help you avoid known problems?

Are standard library concepts used where appropriate?

Has your designer invented a concept that requires a non-standard library or plug-in, when a standard one might have been just as good? Is there an existing implementation that is close enough to a proposed concept that it may be worth adjusting the design to accommodate it?

Are concepts as generic as possible?

Are the concepts in the design needlessly specialized to particular datatypes, or could they be expressed generically? For example, if the design includes a *comment* concept, is the target of a comment any item, or does the design (and worse, the implementation) assume that the target is always a *post* or an *article*?

Can you implement the concepts as separate modules?

If your implementation entangles concepts together, is the lack of modularity really justified? Or are you building up technical debt that will eventually have to be repaid? If you have succeeded in modularizing concepts, are there code dependencies between them that can be eliminated so they can be modified and reused more easily?

Are there complex synchronizations between concepts?

If the product relies on concepts being synchronized in rich ways, does the synchronization produce complexity in the code? If so, might there be a better way to organize it, for example using an event bus or implicit invocation architecture, or by using callbacks and dependency injection?

Do some concept actions involve complex conditionals?

Do some of your concept actions perform elaborate checks on their arguments, or have a complex conditional control flow? If so, this might be a symptom of a troubled concept. Might such an action represent multiple actions (dependent on the arguments presented) within the same concept? Could splitting into several distinct concepts simplify such actions? Is a lack of synchronization between concepts leading to inconsistent states that should not have to be handled?

For Researchers and Software Philosophers

There are many important questions that my evolving theory of concepts cannot yet address. Maybe some of you will be inspired to take up the challenge and help build a more complete theory and method of concept design. With that in mind, here are some open questions.

How should a concept catalog be structured?

A catalog or handbook of concepts would allow designers to codify their knowledge, making it easier for novices to acquire expertise, and would encourage greater reuse of concepts and help designers avoid known pitfalls. How should such a catalog be structured? Should catalogs be domain-specific (e.g., a catalog for social media apps, and a catalog for banking) or should a catalog emphasize concepts that cross domains?

Are there composite concepts?

I have explained how concepts can be composed together, and how sometimes the remedy for an overloaded concept is to factor it into multiple concepts. When a concept is decomposed into smaller concepts, does the larger entity remain as a concept in its own right, and with its own purpose?

Are there different kinds of purpose?

I have given criteria for what makes a good purpose, and a coherence test for identifying when a purpose is composite. But I have neglected some important distinctions regarding the role that a purpose plays. The purpose of a concept, as I explained, motivates its inclusion a design. But inclusion can mean two different things. One is related to the general benefit the concept brings; the other is the particular benefit that it brings in contrast to other concepts that might have been used instead.

For example, both the *label* and *folder* concepts fulfill the purpose of organizing items, and this purpose would motivate including either of them; but only *label* fulfills the purpose of organizing items into overlapping categories. It is not clear to me that this more granular distinction between concepts is even a purpose; perhaps it's a quality that distinguishes concepts with the same purpose.

What issues arise with instantiation of generic concepts?

I've argued that concepts should be stated in a generic form when possible. Doing this allows you to get to the essence of your design, eliminating domain-specific complications that may lead to needlessly unconventional and unfamiliar concepts. When composed, generic concepts are instantiated; when the *trash* concept is composed with the *email* concept, for example, the *items* of the *trash* become *messages*. Is there a way to systematically abstract domain-specific concepts (and purposes) into generic ones?

Instantiation of a generic concept may entail composition with a domain-specific concept. In a restaurant reservation system, the generic *reservation* concept, which knows only about resources, may be composed with a *table* concept that knows about restaurant tables. Exactly such a structure is imposed by the Google Maps reservation API, which requires restaurants to convert a table that can seat four to six people into three distinct abstract resources. Is this its own kind of composition? Are there general principles that lie behind it?

Is action synchronization enough?

Composition of concepts relies entirely (in this book) on action synchronization. Should concepts be allowed to synchronize on state as well? The synergistic composition of *trash* and *folder*, for example, might be expressed as an invariant that relates the items in the trash to the files and folders that are descendants of the trash folder. (See Note 71.)

Can mapping principles be articulated?

Are there general principles for evaluating a mapping? Such principles would presumably rest on well-known principles of user interface design, but would address the connection to concepts more directly. For example, researchers have explored notions of visibility of state (especially with regard to hidden modes), but usually in the simpler setting of a single state machine. What visibility rules might apply to an app composed of concepts?

What role do assumptions about user behavior play in concept design?

Some concepts fulfill their purposes only when users behave in a certain way. The *password* concept, for example, can only provide effective authentication if users pick passwords that are not guessable, remember their passwords, and do not share them. Could such assumptions be expressed as prerequisites of the operational principle?

Can concept implementations be fully modularized?

Concept design suggests a new programming idiom. I explained (in Note 81) why a traditional object-oriented style of programming typically leads to undesirable couplings, often producing a structure in which the dependencies point in exactly the wrong direction. A direct implementation of concepts as modules might produce a more flexible and decoupled codebase (see Note 32). What kind of modularity mechanisms would allow flexible synchronizations and compositions?

Microservice architectures might be a useful basis for concept implementation in which each microservice represents a single concept, and so might be called a "nanoservice" instead. How are nanoservices different from microservices? Can they be synchronized in the manner I've described, without the usual dependencies in which the internals of one service make calls to the API of another?

Can concept design flaws be detected in code?

Ill-formed concepts confuse not only users but programmers too. When experimenting with a conceptual design issue in an application, I have often found the application crashes or exhibits other failures not immediately related to the design issue at hand. I suspect that when concepts are unclear, the code reflects the confusion and defect rates rise; this is certainly my own experience writing code. By mapping files in the codebase to concepts, could source code mining or static analysis exploit this connection? Might conceptual confusions at the design level be predicted by higher defect rates in the code? Might concept design flaws suggest places in the code meriting more careful review?

Can concepts be applied to internal API design?

Concepts are by definition user-facing. But many of the issues that arise when a program makes internal use of a service or API are similar to those faced by users. Might programs in one layer of an implementation stack be regarded, from a concept design point of view, as the "users" of concepts in the lower layers? If so, could concept design principles be applied to code design?

For All of Us

Beyond all these workplace situations, I hope that the ideas of this book will be helpful in that most common scenario—when we're struggling to make sense of yet another application or feature that isn't quite comprehensible. Maybe a little concept analysis will reveal what's going on. At the very least, it might make our daily discussions about the technology we use more grounded and substantive, and help us see more clearly the path to better design.[114]

Acknowledgments

A few years ago, I shared a copy of my book draft with a colleague. It's pretty good, he said—or at least it will be "after a couple of rewrites." I smiled politely, thinking: it's been hard enough writing it once; it will kill me to start again. But in his gentle insistence that the book needed work, he was absolutely right, and I ended up rewriting it three times over. It's still far from perfect, but it's reached the point at which I've explained my ideas as best I can, and it's your turn— my fellow researchers, practitioners and enthusiasts—to join the conversation.

I would never have had the motivation to work on this book for so many years—I first wrote a draft in 2013—had it not been for the ongoing, insightful critiques of friends and colleagues (and their encouragement that it was worth persisting). So many of the good ideas in the book's structure and emphasis come from them. I am, in fact, astonished at the generosity and stamina of those who read the entire book, cover to cover, sometimes more than once. Michael Coblenz, Jimmy Koppel and Michael Shiner gave me copious comments on almost every page; Kathryn Jin, Geoffrey Litt, Rob Miller, Arvind Satyanarayan, Sarah Vu, Hillel Wayne and Pamela Zave gave me excellent expository suggestions; and Jonathan Aldrich, Tom Ball, Amy Ko and Harold Thimbleby not only reviewed the book in detail, but read it again a second time around after I had reshaped the book in response to their wise advice.

The book has been tested for home use too: Akiva Jackson had a slew of brilliant suggestions for improvement that belied his lack of formal education in computer science; Rebecca Jackson, who has become the unofficial editor of everything I write, gave me the benefit of her magic with words; and Rachel Jackson (http://binahdesign.com) shared her exquisite eye for typography and book design.

Readers will likely notice influences on the book that I have failed to appreciate myself, and to all those whose ideas I have taken without credit, I offer apology and gratitude. In addition to the many colleagues that I cite in the endnotes, I'd like to recognize in particular Santiago Perez De Rosso, who was a sounding board for my first ideas on concepts; who built Gitless, the first major experiment in concept design; and whose Déjà Vu system embodied early notions of concept synchronization that we developed together.

My editor, Hallie Stebbins, has given me masterful guidance and advice as I've navigated the path of publication, and has been a staunch advocate for my book from the start; Bhisham Bherwani was a meticulous copy editor; and my production editor, Jenny Wolkowicki, shepherded my book along the way, attending to every detail, and generously tolerated the complexities of dealing with an author who insisted on designing his own book. Kirsten Olson, my coach, inspired me to think of a book project not as the production of an artifact but as an extended conversation and collaboration with an ever-widening circle of colleagues and friends.

The research that underlies this book—as those familiar with today's culture of computer science may guess—was not easy to fund, so I am especially grateful to the SUTD-MIT International Design Centre and its directors John Brisson, Jon Griffith and Chris Magee who stuck with me over five years of support.

I have dedicated the book to my extraordinary parents. My mother, Judy Jackson, has been an inspiration with her many books and projects and her unflagging enthusiasm for all of my activities. My father, Michael Jackson, has taught me so much about software that I can barely tell where his ideas end and mine begin, and I continue to relish our frequent conversations about our field and its history (and the design of elevators and coin-operated zoo turnstiles).

Thank you to you all of you. And finally, to my wife and biggest supporter Claudia Marbach: it's done. I'm so grateful for your wisdom, patience and encouragement, and any day now I'll be ready for that break I've been promising to take for so long...

Daniel Jackson
July 30, 2021

RESOURCES

Explorations & Digressions

How to Read This Book

1. **De Pomiane's Delights.** De Pomiane's *French Cooking in Ten Minutes* [126] is first of all an unexpected source of spiritual inspiration: how many contemporary cookbook writers would insist you cook your meal in ten minutes so that, with only an hour for lunch, you are still left with half an hour to drink your coffee? It also contains some great recipes. My favorite is "Tomatoes Polish Style"—appropriate, as de Pomiane was born into the Polish aristocracy (as Eduard Pozerski in 1875). Melt some butter, add some finely chopped onion, and two plum tomatoes sliced in half face down; cook on high heat for five minutes, turn over, cook five minutes longer, and spoon over two generous tablespoons of sour cream; bring barely to the boil over, take off the heat and serve.

   This is indeed a micromaniac's delight, as it's hard to imagine a tastier dish with fewer ingredients (although I have to admit to adapting it by (a) replacing the onions by shallots, and (b) including some freshly grated nutmeg at the end—and using it as a sauce for pasta in place of de Pomiane's rather too minimal "Noodles Italian Style"). The timing of the recipe also reveals that while de Pomiane might be a micromaniac, he is no pedant.

2. **The importance of details.** "The details are not the details; they make the product." This enduring aphorism by the furniture designers Charles and Ray Eames was written for the script of a 1961 film short called "ECS," about Eames Contract Storage, a furniture system designed for student dormitories. The same spirit is evident in the work of Jony Ive, the famed Apple designer, who ended a charming video about shaping the edges of the Macbook laptop with the self-effacing quip "That's quite obsessive, isn't it?"

   Sorting out the details is rewarding work, but also hard. As Steve Jobs put it [150]: "To design something really well, you have to get it. You have to really grok what it's all about. It takes a passionate commitment to really thoroughly understand something, chew it up, not just quickly swallow it. Most people don't take the time to do that."

Chapter 1: Why I Wrote This Book

3. **The Alloy modeling language**. Alloy is a language and analysis tool for software design. The language itself is a simple but powerful logic based on relations that can be used to model complex data structures and behaviors with declarative constraints (that is, describing the *effects* of the behaviors rather than having to enumerate the steps that produce them). The Alloy Analyzer generates sample scenarios without the user having to write test cases, and can check that the design satisfies properties that the designer formulates—all completely automatically.

   Alloy was inspired by the Z specification language [136] and by the SMV symbolic model checker [23]; its goal was to combine the elegance and succinctness of Z with the analytical power of SMV.

   Alloy's technical innovation is a new analysis that achieves SMV-like automation by limiting the "scope" to finite bounds. For example, an analysis of a network protocol might consider all configurations involving up to five nodes. By compiling into input for a SAT solver, Alloy can promise to cover the *entire* scope, that is all scenarios of that size (which for our network example would involve 32 million cases for the network connectivity graph alone!). This is why Alloy typically finds subtle errors that would elude testing.

   The latest version of Alloy [20] smoothly incorporates the operators of linear temporal logic and includes support for unbounded model checking.

   Alloy has been used in a wide range of applications, including networking, security, and electronic commerce, and has been taught in software and formal methods courses around the world. Alloy is described in length in my 2006 book [66], and more briefly, with applications, in a 2019 magazine article and video [67].

4. **Software design: origins of the idea**. The idea of software design, as advocated in this book—and contrasted, over the following pages, with the idea of design typical in programming, software engineering and user interfaces—has been shaped by many people over many years. But it has never been articulated more forcefully than by Mitchell Kapor in his *A software design manifesto*.

   Kapor's manifesto and Winograd's book. Kapor, the founder of Lotus and designer of its eponymous product, first presented his manifesto at Esther Dyson's PC Forum in 1990, challenging his fellow software executives to recognize the central role of design in software development. The manifesto was printed in Dr. Dobb's Journal (in January 1991), and then appeared a few years later as the anchoring chapter of a seminal book edited by Terry Winograd entitled *Bringing De-*

sign to Software [149]. Winograd, who would later be one of the founders of Stanford's d.school, had convened a group of experts with the goal of defining the term "software design." In the end, his book reflected a diversity of views, but all sharing the conviction that such a discipline existed, albeit in a nascent form; that it had features in common with design in other disciplines; and that it was distinct from both software engineering and user interface design.

Kapor's vision of a new field (and even a new profession) resonated widely, and today everyone pays at least lip service to the idea that software needs to be designed in Kapor's sense of the word. In my view, though, we have yet to build the intellectual foundations on which this field and profession can be based. We have learned much from other design disciplines, and from the experience of creating our own products; and the field of human-computer interaction has flourished. But the particular, distinctive qualities of software design have yet to be articulated.

5. **What is software design?** The term "design" is often used loosely to refer to any activity that involves the creation of an artifact to satisfy some need while meeting some constraints. But in that sense, almost every human endeavor involves design, so the word loses any useful purchase. Some would contrast design to manufacturing, but for software such a distinction is hard to sustain.

I prefer to reserve "design" for the shaping of artifacts judged primarily for their utility—thus separating design from art—and intended for direct human use—distinguishing it from engineering. As Kapor put it [78]: "What makes something a design problem? It's where you stand with a foot in two worlds—the world of technology and the world of people and human purposes—and you try to bring the two together." Engineering focuses instead on cost, performance, resilience, and so on—all concerns that are of great interest to human users but which are usually invisible except when they fail.

Thus an architect, who *designs* a building, aims to create experiences of space and light that delight the building's occupants, and must therefore be intimately aware of their work patterns. The structural engineer, on the other hand, is responsible for ensuring that the building does not fall down in a heavy wind, and that its beams do not rust over time. The designer's analysis is qualitative and by necessity tentative, since the behavior of human users can never be fully predicted; the engineer's analysis is quantitative and definitive, and when it requires assumptions about human users they can usually be reduced to simple numeric measures (such as the maximum number of building occupants, or their average weight).

Software design, the subject of this book, is thus about shaping and structuring the functionality of software to meet the needs of its users. Software engineering, on the other hand, is about structuring the code that delivers this functionality, and subsumes what I refer to in the main text as the "internal design" of software. Some software engineering concerns, such as how fast the software runs or how well it scales, are relevant to the user but mostly invisible, except when limits are encountered. Others, such as maintainability, are relevant only to the developers, and impact the user only to the extent that they impact the cost of development and the feasibility of new features.

The distinction between software design and software engineering was pithily expressed by David Liddle, the leader of the division at Xerox PARC that built the Star workstation: "Software design is the act of determining the user's experience with a piece of software. It has nothing to do with how the code works inside, or how big or small the code is. The designer's task is to specify completely and unambiguously the user's whole experience. That is the key to the whole software industry, but, in most companies, software design does not exist as a visible function—it is done secretly, without a profession, without honor." [149] We'd put less emphasis nowadays on "complete" and "unambiguous" specification, but otherwise his comment might serve as a slogan for this book.

In noting the lack of respect for software design, Liddle echoed Kapor's lament in his manifesto [78]: "Today, the software designer leads a guerrilla existence, formally unrecognized and often unappreciated." Most software, Kapor argued, was "merely engineered" and not designed at all. His remedy was to create a professional discipline of software design, whose practitioners would have a solid grounding in technology but would be distinct from programmers. Their purview would be the very conception of the product, not just the user interface.

It is for just such a designer that my book is aimed. Thirty years since Kapor's manifesto, the title "software designer" is still rare. But the tasks such a designer would undertake are seen as increasingly important, even if they are discharged by people with other titles—whether program managers, architects, UX designers, or programmers themselves.

6. **Programming Knowledge**. In contrast to software design, software engineering (or programming) has a well-established and rigorous body of knowledge. Programming as we know it originated in the late 1950s with Fortran, the first high-level programming language. Within just a few decades, almost all the foundational ideas that we have about programming today were invented: dependencies and decoupling; specifications, interfaces and invariants; abstract types, immutability, and algebraic datatypes; objects, subtypes, generics, and classes; higher-order

functions, closures, and iterators; grammars, parsing, and stream transformations; and so on.

Each of these ideas comes with prescriptive guidance on how to program well, and criteria for distinguishing good programs from bad ones. Three ideas, in particular, seem most significant to me:

Dependences. A *dependence* between two modules arises when the first relies on the second to meet its specification [116], and incurs a liability, in that the first cannot be understood (or used in a new program) without the second. Eliminating dependencies is thus a major goal of program structuring, and is the motivation for many design patterns [44].

Data abstraction. The implementation of a datatype within a module is *representation-independent* [102] if the data structure used to represent instances of the datatype can be altered without having to modify code outside the module, by ensuring that such external code relies only on the behavior of operations of the type [95]. To establish this independence, programmers must not only ensure that clients of the type use it only through its operations, but also that no "representation exposure" occurs in which a reference to an internal structure is leaked to the outside [31].

Invariants. An *invariant* is a property of a program saying that, when observed at certain points (such as before or after particular function calls), the state of the program satisfies some predicate: for example, that a tree is balanced, or that the elements of an array appear in order [40]. (In databases, invariants are called "integrity constraints".) By formulating invariants, the programmer can simplify the task of understanding complex behaviors. Instead of having to consider long histories of events, she can assume that the invariant holds at each prescribed point (so long as the invariant holds at the start and is reestablished after an operation).

This rich theory means that programmers also have an expressive language for talking about programs. A well-trained programmer who overhears a conversation in which one programmer says to another "I'd make the key immutable, because otherwise you risk breaking the hash table's rep invariant" knows exactly what she is talking about and what the issues are. There's no such language, yet, for software design.

7. **Design in software engineering research**. Tim Menzies and his students analyzed the prevalence of different topics in software engineering, with a dataset comprising more than 35,000 papers from top conferences and journals [99]. In 1992, the first year of his study, design was the most popular topic; by 2016, the last year, it ranked near the bottom (eighth in conferences, and not even appearing in his jour-

nal classification). Although this study classified papers in a crude way (checking for a short list of keywords in titles and abstracts), the results match my impressions of a changing field.

Design in human-computer interaction research. Design seems to have had its heyday in the HCI community in the 1980s, with the arrival of the Apple Macintosh in 1984, the emergence of user-centered design, and the publication of Don Norman's *The Design of Everyday Things* [110]. The book, although not ostensibly about software, has had a huge influence on user interface design, most notably through Norman's notions of *affordance* and *mapping*.

Stuart Card, Tom Moran and Allen Newell's landmark book, *The Psychology of Human-Computer Interaction* [26], showed how a model of the user as an information processor with certain parameters (for reaction time, memory capacity, etc.) could reliably predict the efficiency of an interface design, and suggest improvements to it.

In 1989, Jakob Nielsen published his paper on "discount usability," proposing a potent but inexpensive combination of user testing, prototyping and heuristic evaluation for improving the design of user interactions [106]. A year later, a fuller paper with Rolf Molich [107] on heuristic evaluation appeared, leading to the first version of his "10 Usability Heuristics," [108] which are essentially principles of user interface design, and have been followed by several such lists, most notably Bruce Tognazzini's "First Principles of Interaction Design" [143].

Thomas Green developed a list of criteria aimed at the design of programming languages and other notations, which he called "cognitive dimensions of notations" [46, 47]. In fact, his criteria can be applied much more generally to improve usability for any kind of interface. Indeed, several—such as consistency, error-tolerance, mapping and visibility—are common to the other lists mentioned above. Like them, they can be applied synergistically with the principles of concept design. (More on cognitive dimensions in Note 19.)

Design remains a hot topic in human-computer interaction research but it's rare to find a paper that addresses it directly, and from which any concrete design guidance can be extracted. More often, papers explore ethnographic or sociological issues, and don't contribute to a practical body of design knowledge.

For example, last year's session on "Design reflections and methods" in CHI, a flagship HCI conference, included only five papers (out of 748 papers in total), and their topics were: the use of design to study metaphysical ideas in philosophy; modes of reflection for imagining design futures; an agenda for integrating computers with the human body; a feminist/decolonialist analysis of iterative design based on experiences with Aboriginal Australian communities; and the impact of automation on economies in the Global South—all no doubt topics of interest to specialists in those areas, but none of immediate relevance to the vast number of people working on the design of the software that most people use.

For guidance on how to design software, students and practitioners must go to textbooks and professional books instead. I recommend in particular Harold Thimbleby's *Press On* [141], which combines classic HCI principles with a state machine formalism for recording and analyzing designs, and also covers wider social, psychological and ethical concerns.

8. **On verification and its cultural implications**. In the early years of software research, its pioneers—notably Bob Floyd, Edsger Dijkstra and Tony Hoare—introduced the radical idea that the behavior of a program could be precisely specified. With a specification in hand, the difference between acceptable and unacceptable behaviors was no longer subjective, and "bugs"—defects in code that correspond to mismatches between specified and actual program behavior—became a focus of attention.

Dijkstra contrasted the "correctness problem"—whether a program meets its specification—with the "pleasantness problem"—whether the specification is appropriate in the context of use [33]. He noted that the correctness problem can be formulated mathematically, and he thus regarded it as a suitable topic for "scientific" investigation. In contrast, the pleasantness problem, he argued, is "non-scientific" and its pursuit by computer scientists questionable.

This distinction, and the idea of mathematically precise specification, launched the field of program verification, bringing some dignity to our own dismal discipline. Specifications have also been very beneficial in their own right, but perhaps not in the way Dijkstra had expected. It turned out to be impossible to write complete, precise specifications of even the simplest software systems, or at least no easier than writing the code itself. Instead, specifications found their use when applied to much smaller components within a single software system. Such specifications are valuable because they allow bugs to be localized—that is, to identify which component is to blame for an unexpected behavior.

Over the years, the notion of correctness, and the field of verification that grew around it, became a cornerstone of computer science and one of its proudest achievements. From a design perspective, however, its impact has been in some respects pernicious. By separating out the "pleasantness problem" (and giving it such a brilliantly uninspiring name), Dijkstra drew attention away from it, despite its importance. A better name might have been the "design problem" (for creating

189

the specification), contrasted with the "implementation problem" (for building an implementation that meets it).

The implementation problem is undoubtedly a more attractive target for many researchers because it is more clearly defined and more susceptible to incremental progress. Just as the proverbial drunk looks for his lost key under the lamppost because that's where the light is brightest, so software researchers look under many implementation lampposts even though the key to software quality often lies elsewhere. The result of lavishing all our attention on implementation is that we have a field with millions of implementers but few designers, and we often leave the most critical decisions (about what the specification should be) to non-technical people. It's as if the building industry had only civil engineers and managers but no architects [78].

To Dijkstra, bugs were simply defects, and he insisted that it made no sense to talk about the number of bugs in a program: either the program met its specification or it did not. Ironically, the idea of correctness led to an almost exclusive focus on bugs. If the design of the specification is a non-scientific issue, the specification itself cannot matter much, and the primary issues remaining are what bugs are present and how they can be eliminated. This view has come to pervade research in software engineering, where major conferences are now dominated by papers focused on finding and repairing bugs, with barely any discussion of specifications.

Eliminating bugs is not the key to improving software. Of course, software that is riddled with bugs is bad. But software that is supposedly bug-free is not necessarily good. It may still be unusable, unsafe, or insecure.

In short, the argument of this book is that what matters is the fundamental structure of the design, namely the design concepts and their relationship to one another. If you get this structure right, then subsequent development is likely to flow smoothly. If you get it wrong, there's no amount of bug fixing and refactoring (short of starting over) that will produce a reliable, maintainable, and usable system.

9. **Defect elimination and software quality**. The assumption that defect elimination is the key to better software is so widespread that it is rarely questioned (and often not even explicitly articulated). Companies that make software like the idea of defect elimination because it can be applied incrementally, without major disruptions to their development process or to an often shaky codebase. Tool vendors promote it because it helps sell their products. Researchers focus on it because it makes their contributions easier to measure, and because they fear being accused of utopianism if they suggest avoiding defects in the first place.

Defect elimination, however, is not the right focus. Turning a blind eye to egregious defects that can be easily removed is of course unwise. But defects are a symptom and not the cause of low quality. If you don't address the root cause, defects will remain however many you eliminate. And since patches often increase complexity, a software system can become more brittle and unpredictable the more defects you attempt to remove.

A parable about defect elimination. My family lives in a Victorian house, built in the 1880s. It's a beautiful house and it has served us well. But like any old house, it has needed extensive repairs. The biggest challenge has been to keep water out. When we bought the house, water would seep through cracks in the basement walls when there was a heavy rainfall. And after a major snowstorm, water would collect in the soffits behind the gutters, run into the spaces between floors, and eventually drip through the ceilings and walls.

In both cases, the proximate causes were clear, and they have straightforward (although not inexpensive) remedies. To fix the basement leaks, you can repair the cracks in the walls, and spray them with a special waterproofing sealant. You can also install underground drainage to collect the water behind the walls and pump it out. To fix the gutter problem, you can install a layer of rubber ice-and-water shield under the shingles at the edge of the roof, so that runoff from melting snow and ice is directed away from the roof into the gutter.

After many years of experimenting with interventions like these, reading advice online, and talking to contractors, we learned the truth. These kinds of remedies are inadequate (and unnecessary). They don't actually solve the underlying problem; they just mask it.

A better strategy is to identify the real cause and to address that. Take the basement water problem. Water enters your house from the outside, so the best way to prevent it from coming in is to keep it away from the house in the first place. If you grade the land carefully so that it slopes away from the house, and ensure that downspouts from the gutters discharge water far enough away from the foundation, there won't be surplus water pressuring the basement walls.

Dealing with snow on the roof is harder. The problem, it turns out—and this will be familiar to those who live in similar snowy climes—is the dreaded "ice dam." If the roof is warm, snow melts where it meets the roof; it then refreezes when the temperature drops, and a layer of ice forms and eventually grows to cover the eaves. When more melting occurs, this ice prevents water from flowing away into the gutter, and it flows instead under the shingles into the house. The solution, paradoxically, is to keep the roof cold so that the snow melts only when the outside air becomes warm. To achieve this, you can either vent the roof with a gap below the shingles through which cold air can pass, or (if you're stuck with an old house) install insulation on the inside of the roof to keep heat inside the house and away from the roof.

In summary, both of our problems seemed to be caused by visible defects, such as cracks in the basement walls or small openings in the roof. But these defects were just symptoms, and the real causes were design flaws. If a house is designed well (with the land around it well graded, and the roof vented and insulated), water won't come in. Eliminating the visible defects, it turns out, is neither sufficient nor necessary. With enough water or ice buildup, even the best roof or basement wall will fail; and if the design is good, small defects will not even be tested.

10. **Empiricism in software research.** The move in software engineering research away from broader and deeper questions to narrower and more technical ones can be traced back, in my view, to efforts to make the field more empirical. Advocates of empiricism argued (back in the mid-1990s) that the field would garner more respect and make more effective progress if it adopted more "scientific" standards, with experimentation playing a more central role.

The hoped-for benefits did not materialize. Researchers, forced to appease reviewers demanding empirical evidence, turned away from harder questions (especially those whose very formulation is tricky) towards simpler ones that were more amenable to evaluation, and diverted their intellectual efforts to contrived experiments with small sample sizes, often using students as participants to evaluate tools aimed at more expert programmers.

Raising the bar of evidence may have weeded out weaker papers, but it threw the baby out with the bathwater. No longer are submissions judged on the strength and originality of a paper's intellectual arguments and the compellingness of its examples; "results" are now the sole arbiter of acceptability. It's a sobering thought that the most influential papers of the field (such as David Parnas's seminal papers on information hiding and dependencies [115, 116]) could not be published in mainstream conferences today, except in a few more open-minded venues (such as the *Onward!* track of SPLASH).

Similar concerns have been raised in the field of human computer interaction. Saul Greenberg and Bill Buxton have observed that people not only generate research questions that are easier to evaluate, but also often rely on scenarios that are cherry-picked to produce good results, thus proving only that an artifact is usable in *some* context (rather than in all contexts, or in the contexts that matter) [48].

The result, they argue, is that more promising and innovative ideas receive less attention.

In an entertaining and revealing book [13], Laurent Bossavit analyzes a number of folklore beliefs about software—for example, that productivity varies between programmers by a factor of 10, or that prior to agile approaches, software was developed in a strict and inflexible "waterfall" life cycle—and shows them to have no basis. To some, this is evidence in favor of greater emphasis on empirical data. But as Bossavit notes, the real problem lies not in the original papers that are cited in support of such beliefs, but rather in the game of "telephone" that followed, in which a paper's content and message were progressively degraded and misrepresented. The problem therefore is not that the original papers lack compelling data, but that they are not read critically (or indeed at all).

Of course, not *all* empirical studies are suspect. My objections are to an unthinking preference for empirical evaluation over other forms of analysis (inspired my misleading analogies to the hard sciences), and to the assumption that all ideas benefit from quantitative assessments. When targeted appropriately and conducted imaginatively, empirical investigations can of course be revealing and valuable.

In the field of programming languages, for example, it has long been recognized that programming is as much about communicating with other programmers as it is about conveying information to a compiler. But only recently have researchers begun to embrace the idea that programming languages are a human tool, and that questions of usability are fundamental to their design [28].

Jonathan Aldrich, a noted programming languages researcher who is a leader in this new area, has balanced his technical contributions to programming language semantics and theory with careful investigations of the impact of language features on users. A study led by one of his students, for example, analyzed a corpus of open-source programs, confirming that they indeed would have benefited from structural subtyping, a language feature popular in research languages but rarely deployed in practice [98]; another looked at the prevalence of object protocols in code, finding that enough modules used them to make protocol-checking tools worthwhile [9].

11. **How concepts aid design thinking**. The emergence of design thinking as a popular movement has done much to elevate the role of design in our society and set higher expectations for the design quality of everyday artifacts. It has also inspired many people to think in a more open and creative way about all aspects of their lives, by questioning assumptions, imagining radically new solutions and reevaluating needs.

The "content-free" nature of design thinking—that the processes it advocates are independent of any domain-specific design principles, language, or strategies—makes it a good match for concept design, which can provide the substance for design thinking in the context of software. In the needfinding phase, the idea of concept purposes can be used to refine and structure needs; in divergent, ideation phases, existing concepts can be drawn on, enabling more ambitious and yet more lightweight exploration; and in convergent phases, concepts provide language for recording designs and criteria for evaluation.

Perhaps most importantly, concepts allow a design exploration to be factored into separate explorations, one per concept (or one per purpose that a nascent concept is intended to address), which might proceed sequentially or in parallel. A design thinking project can be sunk by too large a scope that leads, in divergent phases, to an array of design ideas that is too large and unstructured to be amenable to convergence. It would make little sense, for example, to pursue the design of a healthcare information system as a design thinking problem. By identifying individual concepts and their purposes—for example, the problem of diagnosing conditions, or triaging patients, or scheduling appointments—the design effort can be given some structure, allowing design thinking to be applied in a more granular and focused way.

Some qualms about design thinking. Part of the appeal of design thinking has been its accessibility, and the inclusive message that it sends to all members of an organization encouraging them to engage in design activities. Broadening participation in design is surely a good idea. Time and again, designers have found that engaging with users and community members results in better designs. For Christopher Alexander, patterns are valuable precisely because, by embodying experience and design wisdom, they make it possible for people without experience to take advantage of it. Accumulated design expertise, in other words, is the basis for democratization.

But this enthusiasm has a downside, and has led to some misapprehensions about design. One gets the impression from the way many design thinking books talk that design is an easy and fun activity in which radically new forms are conjured out of thin air, and that a grounding in a particular design discipline is not only unnecessary but may be an impediment to fresh thinking.

This misrepresents design in several key respects. First, the best designers don't work in a vacuum. They are steeped in knowledge and experience of prior designs, which they draw on as they imagine new designs. Second, most designs aren't radical new forms, but are subtle modifications of existing forms; the genius of design is more often in the details, and in how different design elements are reconciled, rather than in the novelty of the design as a whole. And third, each design discipline really does call for its own sensibilities. Natasha Jen raised similar concerns about design thinking and its deemphasizing of the role of design criticism in her talk "Design Thinking Is Bullsh*t" (and pillories the ubiquity of 3M Post-its as the design tool du jour) [74].

Design in other domains. There is undeniably a "design mindset" that applies across domains, and my understanding of design has been enriched by many books that are not about software. My favorites are those describing and diagnosing failures, such as *Why Buildings Fall Down* by Mario Salvadori [93]; *Why Buildings Stand Up* by Mattys Levy and Salvadori [133]; Henry Petroski's *To Engineer is Human* [121]; and Charles Perrow's *Normal Accidents* [120].

I have wondered if it would be possible to write such a book about software failures, but I suspect not, for a simple reason. These books are appealing because they tell compelling stories of designs gone wrong: a plan that seemed perfect but then failed spectacularly because of an unwarranted assumption, or a flaw in execution. When failures of a similar magnitude happen to software—for example, a security breach that leaks the personal records of millions of people—the diagnosis is invariably that no reasonable safeguards were ever included in the first place. Without any design rationale for success, there was no reason to imagine that the failure would *not* occur, and there is thus no design story to tell. A story about why companies are not incentivized to design software more carefully can still be told, but it's about commerce and risk instead.

Sources of inspiration. Seeking inspiration on design from other disciplines, I have been influenced most by Michael Polanyi's notion of the "operational principle" [125] (introduced to me by Michael Jackson [72]), Nam Suh's independence axiom (from his Axiomatic Design theory in mechanical engineering [137]), and, more amorphously but no less importantly, Christopher Alexander's ideas about form, context and fit, and about the role of patterns in design [3, 4, 5].

Books on typography and graphic design offer an enviable collection of design ideas and design theories, and are full of principles, patterns and illustrative design examples, suggesting a model for how to write about design. Traditional texts on typography frequently give prescriptive design guidance; Jan Tschichold's *The Form of the Book* [144], for example, gives a systematic treatment of page layout and advice on how to use different ratios in shaping the page and the text block. Most impressive is Robert Bringhurst's *The Elements of Typographic Style* [16], not only for the quality and quantity of the design advice that it contains, but also for the beautiful demonstration the book itself provides of how successful typographic design can be. Even if the design ideas in these books don't carry over to software, they inspire in the recognition that design principles are a good thing, and amplify creativity rather than constraining it.

Software development has long looked to other domains for inspiration. The best-known example of borrowing from older fields is the idea of reusable components or interchangeable parts, which goes back to Eli Whitney's demonstration to President John Adams in 1801 of a musket assembled from prebuilt components. The demonstration was later proven to have been faked: Whitney had marked the components and they weren't fully interchangeable. Nevertheless, this moment is often cited as a technological turning point from handcrafting to industrial production. The value of components in software was first articulated by Doug McIlroy at the 1968 NATO conference that launched the field of software engineering [101], and is important in concept design too, with concepts themselves as the reusable components.

12. **Formal specification and design**. In the 1970s and 1980s, researchers developed a slew of languages for specifying the behavior of software systems using mathematical logic. Some of these—the so-called "model-based" languages such as Z [136], VDM [75], and B [1]—abstracted away the low-level details of the software implementation by describing the behavior in terms of actions over abstract states (comprising sets and relations, rather than the objects, classes, and linked lists of code). Others—the so-called "algebraic" languages such as OBJ [45] and Larch [51]— went further, and described the behavior without any state at all, using axioms relating observers (actions that reported on the hidden state) to mutators (actions that changed the state).

For the most part, these languages were motivated by the conviction that software quality means correctness: that the behavior of a program conforms to its specification. Clearly, without a precise specification, correctness cannot even be judged, let alone effectively pursued.

And yet, as researchers began to write formal specifications, they discovered that the very activity of writing them revealed inconsistencies and confusions in the intended behavior. Far from being a simple matter of recording expectations that were already clear, the act of constructing specifications was a powerful design activity in which many of the most critical decisions about a system were made. This is evident, for example, in a book of elegant case studies in the Z language [55], and was articulated as an explicit goal in a beautiful demonstration of the use of Larch to design the essential properties of a window manager [50]. In the field of human-computer interaction, Harold Thimbleby in particular has explored the benefits of formal specification, with a chapter in an early book [139] showing how algebraic properties can be applied to user interface actions, and later edited a collection of papers on the role of formal methods in HCI more generally [54].

My own Alloy language (see Note 3) was, from the start, intended for design exploration. It was still a surprise to us to discover as we began to use it that the most useful analyses were often not assertion checks, in which a design is tested against expected properties, but simulations, in which arbitrary scenarios are generated, prompting the designer to consider unanticipated (and often pathological) cases.

13. **On simplicity and clarity.** Pondering Zoom's dominance in the video conferencing market (suddenly enlarged by the new working conditions of the COVID-19 pandemic), tech writer Shira Ovide attributes the company's success to the fact that its software "just works." As she explains, "Being the first or even the best at something may not matter." Instead, "Simplicity is the overlooked secret to success." But she recognizes that designing for simplicity is no easy task: "It's the deceptively difficult ticket to riches." [114]

Ovide was unwittingly echoing the views of two of the most famous computer scientists, Tony Hoare and Edsger Dijkstra, both avid proponents of simplicity. Hoare's remarks on simplicity in his Turing Award Lecture [57] have become perhaps the best known quotes in software engineering. Both criticized excessive complexity in programming language design. The first, about Algol 68, lamented the rejection of a proposal for a simpler language: "I conclude that there are two ways of constructing a software design: One way is to make it so simple that there are *obviously* no deficiencies and the other way is to make it so complicated that there are no *obvious* deficiencies."

The second, about PL/1, noted that simplicity is elusive even when (or perhaps especially when) the resources needed to achieve it are readily available: "At first I hoped that such a technically unsound project would collapse but I soon realized it was doomed to success. Almost anything in software can be implemented, sold, and even used given enough determination. There is nothing a mere scientist can say that will stand against the flood of a hundred million dollars. But there is one quality that cannot be purchased in this way—and that is reliability. The price of reliability is the pursuit of the utmost simplicity. It is a price which the very rich find most hard to pay."

And here's Dijkstra [36]: "The opportunity for simplification is very encouraging, because in all examples that come to mind the simple and elegant systems tend to be easier and faster to design and get right, more efficient in execution, and much more reliable than the more contrived contraptions that have to be debugged into some degree of acceptability."

Hoare and Dijkstra were concerned more with software engineering than software design (although Dijkstra abhorred the former term and never used it himself), so naturally they see the benefit of simplicity primarily for reliability. The benefits in the realm of software design seem even greater, since users have less tolerance for complexity than programmers. My belief is that designing software with concepts not only results in a better user experience, but also in more reliable software, because clarity in the design leads to clarity in the code.

In Zoom's case, simplicity may not be the only explanation of the company's success: the quality of the video alone partly explains its dominance over competitors such as Skype and Google Hangouts. And, as we will see later, the Zoom app suffers from several conceptual design problems.

14. **The origins of conceptual models**. In a historic paper [25], Stuart Card and Tom Moran summarize their work over the years at Xerox PARC. Although most of the paper describes their pioneering cognitive model (the "human information processor") and its application to designing the primarily physical aspects of user interfaces, they also discuss the role of mental models, and advocate the view that I have adopted that the mental model is not accidental but is to be explicitly *constructed* by the designer through invention of an appropriate conceptual model. As they put it: "It is clear that users attempt to make sense—by building mental models—of the behavior of a system as they use it. If a simple model is not explicitly or implicitly provided, users formulate their own myths about how the system works ... [I]f the user is to understand the system, the system has to be designed with an explicit conceptual model that is easy enough for the user to learn. We call this the intended user's model, because it is the model the designer intends the user to learn."

In the preface of the 2002 edition of *Design of Everyday Things* [110], Don Norman lists the most important design principles in his book. First on his list is the *conceptual model*, which he illustrates with the example of a thermostat, and how a user lacking a correct model may set the temperature higher in the vain hope of getting warmer faster. (The other principles on his top-four list are giving feedback to users, imposing constraints to prevent errors, and signaling affordances.) To Norman, however, the conceptual model principle is mainly about communication: that the appearance of a device should convey its conceptual model (for more on this, see the discussion of Norman's refrigerator in Note 54). My view in this book is closer to Card and Moran's: that the shaping of the conceptual model is itself the primary design challenge, and the problem of conveying it (or, in my terminology, *mapping* the concepts to a concrete user interface) is secondary.

Conceptual models of APIs. Just as a user needs a sound conceptual model to operate software, so a programmer needs one to incorporate another programmer's code through an API (application programming interface). One study [81] of programmer comprehension of API documentation found that programmers without a basic grasp of the API's concepts struggled even to formulate search queries, or to assess the relevance of the content they found, making effective usage of the API nearly impossible.

Fred Brooks and conceptual integrity. In 1975, Fred Brooks wrote *Mythical Man Month* [17], based on his experience managing the OS/360 project at IBM. The book became a classic, and has been extremely influential. One of its key ideas was "conceptual integrity," which Brooks claimed was "the most important consideration in system design." In an afterword to the 1995 anniversary edition, he reflected on the views he'd expressed in the original edition, notably retracting his opposition to David Parnas's idea of information hiding. In this regard though, his opinion was unchanged: "I am more convinced than ever. Conceptual integrity is central to product quality."

Brooks expressed a similar view in his influential *No Silver Bullet* paper [18], in which he divided the challenges of software development into essence—"the difficulties inherent in the nature of software"—and accident—"those difficulties that today attend its production but are not inherent"—and located the essence in the concepts underlying the software: "The essence of a software entity is a construct of interlocking concepts: data sets, relationships among data items, algorithms, and invocations of functions. This essence is abstract in that such a conceptual construct is the same under many different representations." Furthermore, he argued that developing the conceptual structure is the greater challenge: "I believe the hard part of building software to be the specification, design, and testing of this conceptual construct, not the labor of representing it and testing the fidelity of the representation."

For Brooks, conceptual integrity requires that the design of the entire system emerge from a single mind. Consistent with this view, in his latest book, *The Design of Design* [19], he defines "style" as a set of different repeated microdecisions, each made the same way whenever it arises. In contrast, in the book that Brooks coauthored on computer architecture [12], the notion of conceptual integrity is given a brief definition as comprising three essential properties: orthogonality, propriety and generality. Brooks himself seems not to have developed these ideas further, although they have been inspirational to many others (including the author of this book).

The field of conceptual modeling. There is in fact an entire subfield of computer science called "conceptual modeling," but its focus is a different kind of conceptual model. Here, the model captures entities and relationships in the real world; the term "conceptual" is used to distinguish the representation inside a computer from the external reality. Conceptual models of this sort are used for classical AI reasoning (e.g., in a robot planner), or for defining databases in applications (such as payroll, or indeed almost any kind of information system) that maintain data about the real world. Another, more recent, focus has been to capture the structure of knowledge in the World Wide Web.

At its heart, conceptual modeling is a descriptive endeavor. As John Mylopoulos, one of the leaders of the field explains: "Conceptual modelling is the activity of formally describing some aspects of the physical and social world around us for purposes of understanding and communication" [105]. In contrast, this book is about design and invention; description, while vital, is not an end in itself.

Conceptual models themselves are thus usually data models (also called semantic ontologies) that express the fundamental elements of a problem domain and their relationships. In database development, such data models are contrasted with database schemas, which specify not only these problem domain aspects but also how they are represented in the database (for example, as a collection of tables). The most popular data model used in conceptual modeling is the entity-relationship (ER) model. The ER model itself was developed by Peter Chen in 1976 [27] as a stepping stone in database design, and was hugely influential. Other models (for example, the Semantic Data Model [52]) offered richer features, but few of these were taken up, with one notable exception: the ability to specify that one entity is a subset of another. With this feature added, the model is known as the "extended entity-relationship model," and is the basis for the notation used in the Unified Modeling Language.

Surprisingly, the field does not seem to have reached consensus on what exactly a concept is; the word seems to be used only in its adjectival form. As a recent paper [117] puts it rather stridently: "The conceptual modelling community not only has no clear, general agreement on what its models model, it also has no clear picture of what the available options and their implications are. One common claim is that models represent concepts, but there is no clear articulation of what the concepts are."

Most researchers would probably point to the entities in a conceptual model as comprising the "concepts." This would align the concepts of conceptual models with the concepts of formal concept analysis, which organizes concepts into a lattice (essentially a taxonomy in which a concept can have more than one parent). In this view, the associations of a conceptual model express relationships between concepts.

But a distinction between concepts and relationships is not tenable, and depends on how the model is constructed. A conceptual model of a restaurant reservation system, for example, might have an entity called *reservation*, associated with other entities such as the table being reserved and the customer making the reservation. On the other hand, reservations might instead be encoded as an association between customers and tables. One reason to prefer treating a reservation as an entity is that attributes, such as the date of the reservation, can be added. But some modeling languages (including the original ER model) allow associations to have attributes too, so that consideration evaporates. Either way, it seems clear that a definition of concept that distinguishes entities from relationships is on shaky ground.

A larger problem with this approach (identifying elements of the data model as concepts) is a lack of structure in the model, without which concepts proliferate. If every entity or relationship in a conceptual model is a concept, then presumably the start and end times of a reservation are concepts too. To find a practical notion of concept in a conceptual model, I believe we need to break the model into pieces, grouping multiple entities and relationships together. This is what my concepts do: all the elements that support the booking of reservations become part of the *reservation* concept, with its own localized data model.

Fowler's analysis patterns. My concepts are closer to (and were influenced by) Martin Fowler's "analysis patterns" [42], which are small, reusable conceptual models. An important difference is that concepts are primarily behavioral; their structure supports the behavior by identifying what needs to be remembered (in the state of the running concept) to produce the behavior. Fowler brings in behavior by moving towards the code, and showing methods associated with classes. As we will see, this isn't necessary, and behavior can be specified in an implementation-independent way.

Domains, data models and domain-driven design. A related idea is "domain modeling," which advocates using a model of the problem domain as the basis for developing a software system. An early approach to software development that relied on explicit modeling of the problem domain is Michael Jackson's JSD [68]. In JSD, each entity of the problem domain is modeled as a context-free grammar capturing the possible sequences of events in the life of the entity. System functions are defined in terms of the model, and an implementation is obtained by a systematic transformation of the model and system functions.

Object-oriented approaches also advocated modeling of the problem domain, although objects turned out to be too polluted with implementation decisions to be useful modeling constructs. The Object Modeling Technique [131] found an elegant way to square an object-based implementation with more faithful modeling: an entity-relationship domain model is constructed first, and then (in a similar spirit to JSD) is transformed into object structures.

A popular book by Eric Evans entitled *Domain Driven Design* [38] has brought domain models to practitioners, breathing new life into an old idea. In addition to domain modeling, the book renews other important but neglected ideas, such as the distinction between "entity" and "value" objects [96], the idea of a layered architecture in which a lower layer provides a language for a higher layer [32], and the importance for teams of using common terminology [151]. One key respect in which the book extends traditional approaches to domain modeling is the idea of *bounded context*: that different (and even incompatible) domain models may be needed for different areas of functionality and parts of an organization. Concept design takes this a step further, with each concept holding its relevant part of the data model.

A similar decomposition of the problem domain into distinct subdomains plays a role in Michael Jackson's problem frames [70], which structure requirements as archetypal patterns that capture the relationships between phenomena within the system being built and phenomena in the domains with which the system interacts. In his most recent work [71], Jackson describes a system as a collection of "triplets." Each triplet consists of a machine (a program that executes on a computer); a portion of the governed world that the machine interacts with; and a behavior that results from the interaction. His approach is notable because it allows each machine to work with a *different* model of the governed world, recognizing that for different kinds of behaviors, different aspects of the world will be relevant (and different approximations will be acceptable). His machines are less granular than concepts, but like concepts they have their own data model and dynamics.

Other researchers have explored domain modeling as the basis for requirements: Dines Bjørner, in particular, placed such emphasis on domain modeling that he referred to it as "domain engineering" [10].

Concepts in computer system design. The field of computer science known as "systems," which is focused primarily on the design of infrastructural components (such as networks, file systems, etc.), tends to be case-based, with more focus on breakthrough ideas than on systematizing the variety of possible designs. Notable exceptions are the textbook by Jerry Saltzer and Frans Kaashoek [132], which identifies design themes that are close in some respects to my concepts (and whose chapter on naming, in particular, influenced my ideas), and the course notes on system design by Butler Lampson [86], which shows how the behavior of complex components (such as distributed memory) can be characterized by precise (and often surprisingly weak) specifications.

Chapter 2: Discovering Concepts

15. **The Unix origins of Dropbox's folder concept.** Dropbox adopted the concept of *folder* from the Unix operating system, in which a folder is called a *directory*. This design has many elegant aspects. In particular, since the names of files and directories are *not* treated as metadata but are simply contained within directory entries, there is no need for any additional structure in the file system for maintaining this information; directories can be represented with data blocks just like files, albeit with a special interpretation.

Allowing a file or directory to have more than one parent, a feature of Dropbox adopted from Unix (and essential for expressing sharing), is powerful but even in the single-user Unix setting brings some nasty complications. Deletion of a file does not simply eliminate it, but rather deletes a directory entry—and the file might still be linked through another directory. Consequently, reclaiming storage requires a form of garbage collection to identify inaccessible files.

From the user's perspective, the possibility of multiple parents produces at least three other surprises. First, novice Unix users, expecting a folder to "contain" other files and folders (rather than just containing named links), look in vain to find an option for the directory-listing command that tells you how big a directory is— that is, how much disk space it occupies. Such an option does not exist, arguably for good reason (since a file can have two parent folders, so it's not clear how the file's space consumption should be assigned). Instead you need to use a different command (called *du*, for "disk usage") which, in classic Unix style, will generate by default a report of sizes of all reachable directories without specifying the unit of

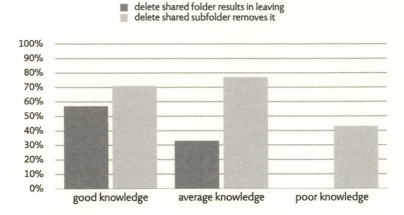

FIG. E.1 *Results of a survey testing understanding of Dropbox amongst computer science undergraduates at* MIT: *the bars show the proportion of correct answers to two questions.*

measurement! Needless to say, only computer scientists would tolerate such un-usable software, so when the Unix file system was adopted by Apple for the Mac-intosh, this was all hidden away, and the Finder displays folder sizes as expected.

A second surprise, which is harder to work around, and becomes a serious prob-lem, is that changing the name of a file is indistinguishable from deleting it from a directory and then adding it back again under the new name. Tools built on top of Unix for synchronizing files across machines therefore can't tell when a file is re-named that it's the same file, and not a new one that happens to have the same con-tents. This makes it impossible to reliably track the history of changes to a file. If a new file has the same contents as an old file with a different name, was the file just renamed or was the old file deleted and a new one created that happens to have the same contents? The Git version control system, for example, guesses that the file was renamed if the file is large, on the grounds that it's less likely in that case that you'd create a new file with identical contents! Dropbox doesn't suffer from this problem because, unlike Git, it can see the action in which the user renames the file, and even if the file system is Unix-based, it can interpret that renaming in its own, non-Unixy, way.

A third surprise is that someone who has no permission to read or write a file may still be able to move it and change its name, since those actions are applied to a directory containing it, and not to the file itself.

This *folder* concept is perhaps one of the oldest concepts in computer science. It was actually invented in the late 1960s—in the predecessor to the lab I now work

FIG. E.2 *Warning message displayed by Dropbox app after a shared folder has been deleted.*

in at MIT—as part of an operating system called Multics. The flexibility of having more than one parent was only partly present in Multics, which had two distinct ways in which one directory could refer to another. The main references were called "branches," and they always formed a tree. The others were called "links" and were unconstrained. Unix eliminated the distinction, using links (properly called "hard links") in both cases, so that when a file (or directory) is accessible from two or more directories, the situation is completely symmetrical.

16. **An empirical study of Dropbox users**. It's not just the technologically unsophisticated who get confused by these concepts. My student Kelly Zhang surveyed about 50 computer science undergraduates at MIT, asking them first to rate their understanding of Dropbox, and then to predict the effect of deleting a folder in two cases. The results are shown in Figure E.1. The effect of deleting a top-level shared folder (such as *Bella Party*) was understood by less than 60% of those who said they had "good knowledge" of Dropbox, by less than 40% of those with "average knowledge," and by none of those with "poor knowledge." The effect of deleting a folder contained in a shared folder (such as *Bella Plan*) was predicted correctly by about 70% of those with good knowledge, almost 80% of those with average knowledge, and only just over 40% of those with poor knowledge. In short, if you delegated control of your Dropbox to an MIT computer science undergraduate, you wouldn't do much better than if you just tossed a coin to decide what to do.

17. **Dropbox mitigations**. To Dropbox's credit, even if you delete a file by removing the mirrored copy on your desktop (rather than deleting the cloud file more directly through your browser), a warning akin to the browser warning (Figure 2.3) is displayed, as shown in Figure E.2. But this warning is shown *after*, not before, the file or folder has been deleted, so it's easy to miss.

Dropbox's design also includes a variant of the *trash* concept, explained in Chapter 4, so that files and folders that are deleted are actually moved to a special temporary location and can be retrieved until 30 days have passed, after which they are lost forever.

18. **Time to sue Dropbox?** Given this chapter's rather critical analysis of Dropbox's design, you may wonder if I'm advocating a class action suit against the company. Far from it. When the Apple Macintosh appeared, one reviewer noted that it was the first user interface design that was *good* enough to be criticized. Previously, the user interfaces of most computer systems were so inconsistent and arbitrary that no coherent critique was possible. Throughout this book, I've picked examples from products made by leading companies, both because they are likely to be familiar to readers, and because they represent the best and not the worst examples of software design. It would be easy to critique a straw man—an incoherent design by an unknown company. But by focusing on major products, I hope to convince you that concept design has value even for companies that seem to have nearly unlimited resources and that employ the most talented designers and engineers. If concept design can reveal serious flaws even in their products, imagine how much more useful it can be for the long tail of smaller and less well resourced companies.

19. **Levels of UX design for software.** In the late 1970s, James Foley and Andries van Dam identified four levels: lexical, syntactic, semantic and conceptual [41]. These levels reflected the structure of the implementation; indeed, the first three correspond exactly to the classic compiler structure that had emerged in the prior decade. In contrast, my levels reflect design concerns. A red button, for example, would sit at their lexical level, but in my analysis might invoke design questions at both the physical level (will color-blind users see it clearly?) and the linguistic level (does red mean stop?). The behavior of the button would be placed at their semantic level, but where it goes in my scheme would depend on whether the pressing of the button is conceptually significant (a moderator rejecting a post, for example) or not (aborting a slow query, say). And despite similar terminology, their conceptual level is different from mine. Theirs is concerned with the user's mental model, and focuses on goals rather than details of behavior; mine embodies the essential behavior that comprises the shared understanding of user and designer alike.

Tom Moran proposed a three-level scheme that is much closer to mine, but like Foley and van Dam's, also more influenced by implementation structure [103]. At the bottom, he places the *physical component*, which handles devices and user interface layouts; in the middle, the *communication component*, which involves interactions (such as key presses) and the syntax of the command language; and at the top, the *conceptual component*, which comprises tasks and their meanings. What is physical to Moran depends on the computer; in my scheme it depends on the perceptions of the human user. Many design aspects are like the redness of the button and have both physical and linguistic aspects: just consider how links are signaled on web pages by underlining text or coloring it blue. On the other hand, Moran's conceptual level—unlike Foley and van Dam's—is similar to mine. He divides it into two sublevels, the task level (which expresses the goals of users in terms of the tasks they want to perform) and the semantic level, which comprises a collection of entities with associated operations. The innovation of this book is thus not in identifying the conceptual level but in giving shape to it. As we'll see, we can find useful structure in the conceptual level by going beyond entities and operations.

Bill Buxton also criticized Foley and van Dam, preferring Moran's levels [24]. He argued for more explicit consideration of "pragmatics," by which he meant the aspects of design that correspond to basic human interactions. He noted that our ability to perform mental chunking, for example, may determine how complex a command grammar can be tolerated. In the schemes of both Moran and Foley/van Dam, syntax occupies the middle level. In my scheme, in line with Buxton's view, this particular question would be placed at the physical level, because—like perceptual fusion—it belongs more to the user's cognitive rather than the linguistic characteristics.

Ignoring UX levels. Some authors treat the user's experience of a system's interface as a single, integrated entity, and do not distinguish levels. Thomas Green, for example, in his first paper on "cognitive dimensions of notations" [46] presents the slogan "system = notation + environment" precisely to suggest that the underlying semantics of the notation and the tool in which it is embedded are inseparable.

To the extent that a designer can mitigate flaws in the notation proper by more elaborate tool support (for example, functions to display otherwise hidden dependencies), this observation is helpful. But it runs counter to the fundamental premise of this book: namely that separation of levels allows for greater clarity and effectiveness in design, and that deep flaws at the conceptual level cannot be remedied by linguistic and physical band-aids. For designing a notation, this would mean at the very least distinguishing between semantics, abstract syntax and concrete syntax, which (surprisingly) Green does not do.

Finding semantics in usability. If you detected in my three-level classification a desire to emphasize semantics over syntax, you'd be correct. As one colleague quipped, parodying our field's tendency to favor one over the other: in computer science, the term "semantic" usually just means "better." This is a little unfair, and I certainly recognize the critical importance of physical and linguistic design, and especially the way in which subtle choices of type, color, layout and language can impact the emotional experience of the user. But it would not be too reductionist to say that this book is, to a large degree, an attempt to find semantics in usability.

20. **The dystopia of Terminal World.** David Rose [130] describes a dystopia that he calls Terminal World in which every interaction we have with a physical object is through a glass slab and glowing pixels. Bret Victor has also railed against the limitations of devices that use the human hand for input but allow only pointing and sliding [146].

    The concerns of the physical level would apply even in Terminal World, but they become more interesting when physical interactions are enriched. Although phones and computers seem (largely due to Apple) to be carrying us inexorably towards Rose's dystopia, in other areas of design, resistance to the universal interface seems to be growing. Users are tired of menus and clicking, and prefer more tactile experiences. Some camera manufacturers (notably Fujifilm and Leica) have preserved classic mechanical designs in their digital cameras, with enough knobs and dials that adjustments can be made without looking at a screen. For many photographers, this feature alone makes such cameras preferable to others with more elaborate features.

21. **Fitts's Law and "physical" capabilities more generally.** Another example of a UX design principle grounded in the physical capabilities of the human users is Fitts's Law, which predicts the amount of time it takes to move a pointing device to a target. Simplifying, the time varies inversely with the width of the target, because if the target is small, the user has to slow down and might have to move back and forth to find exactly the right position.

    A classic application of Fitts's Law demonstrates the superiority of the menu design in macOS to that of Windows. In the macOS design, the menu items at the top of the desktop present an effectively infinite width: the mouse sticks as you attempt to move it past the desktop boundary. Consequently, opening a menu is faster and easier than in Windows.

    Note that the physical level of design accounts for "physical" capabilities of users in the broadest sense; it includes cognitive aspects such as the user's memory capacity. The classic work on the physical level of user interface design is *The Psychology of Human-Computer Interaction* [26], which bases a theory of interface design on an explicit model of the human user as an information processor, with parameters for reaction time, memory capacity, etc. The book actually derives Fitts's Law from this model.

22. **Risks of linguistic misinterpretation.** The icon I've used in Figure 2.5 to represent the linguistic level is a British traffic sign warning drivers of road works ahead, affectionately known as "man having trouble opening umbrella," illustrating the risks of unexpected interpretations at this level.

23. **Redundant functionality, bloat and discoverability.** It's rare for apps to lack features that their users desperately need. When this happens, the users complain (or go elsewhere), and developers respond accordingly. Having too many features, or features that are too complex, is a more real concern. A mantra of agile programming is to build "The Simplest Thing That Works." Another reminds you when considering complex functionality that "You Aren't Going to Need It" (abbreviated YAGNI). The wisdom in these slogans reflects the painful experience of many software teams that the most ambitious, and seemingly essential, features often consume the lion's share of effort and end up not being used at all. On the other hand, the complaint that some very successful apps are "bloated" fails to recognize that while it may be true that only 20% of an app's functionality is essential, that 20% may be different for different users.

    The flip side of YAGNI is the recognition that for any particular purpose, you *are* likely to need the complexity that others have found necessary in fulfilling it. So don't try to build authentication and imagine that you don't need to let users reset their forgotten passwords; or build a shopping cart in which you can't change the count of each item at the last minute; or design almost any professional app without offering presets to store and recall complex settings.

    The only way to know what users are likely to need is through experience. Concepts help because their design embodies experience accumulated across many applications in many different contexts. So when you're designing your authentication mechanism, for example, you should select from established concepts and not be tempted to roll your own. That way, you won't stray too far in either direction (of excessive complexity or inadequate functionality), and you won't introduce a critical security vulnerability.

    Discoverability is a real issue. My favorite example of a concept that took me years to discover: Apple Keynote's *object list view* (introduced in Keynote 7.1, March 2017), which shows the objects on a slide in a tree and allows you to make cross-cutting selections, so you can alter formatting, animations, and so on arbitrarily, without the constraints imposed by layers and groups. PowerPoint has a similar concept called *selection pane* which was introduced earlier (in the 2007 version) but apparently also went unnoticed for years by many users (as indicated by a rhapsodic review in a blog post dated 2013).

I suspect that an increasingly common reason that some features go unnoticed is that they are not accessible through keyboard or menu commands. Menus, despite being old-fashioned and a source of visual clutter, at least have the benefit that they can be easily scanned for available actions. As interfaces have become increasingly visual and less textual, they have become harder to explore. I wonder, for example, how many users of Apple Preview (a PDF viewing and editing app) know that PDF documents can be merged by dragging thumbnails from one window to another, or that an Apple Keynote presentation can be made from a set of photos by dragging the selected file icons to the slide navigator window.

24. **Conceptual models of varying degrees of sophistication**. Sometimes, the same app or system can be viewed at varying degrees of sophistication. This happens less often than you might imagine, because in well designed apps the designer's concept is only as complex as it needs to be to support the intended purpose, and a user who fails to grasp the concept won't be able to use it effectively. A concept may have behavioral details that a user is not familiar with, but these are rarely an impediment to usability, and can often be learned on the fly.

Such varying degrees of conceptual understanding tend to appear when a concept represents a mechanism that is hidden from most users, but may become visible either if a user's activities extend into a new realm, or if failures occur. Someone browsing the web, for example, may imagine that *amazon.com* is the name of a machine that is owned by Amazon, and that typing this name into the bar at the top of the browser, and subsequently interacting with the pages returned, causes the browser to contact this machine, sending it queries and receiving responses back. While oversimplified, this view is sufficient to use a browser effectively most of the time, and might be formalized as a *web service* concept. In particular, it allows a user to understand who might have access to data that is entered into the browser.

In contrast, a user who doesn't even have a model that distinguishes the servers owned by different companies won't be able to understand why entering private data is safer on some sites than others, and arguably has a conceptual model that is simply incorrect. In a survey of users' models of the Internet, one user responded to a question about where data goes as follows: "I think it goes everywhere. Information just goes, we'll say like the Earth. I think everybody has access." [77]

A richer conceptual model would recognize that Internet servers do not have intrinsic, symbolic names, and that a name appearing in a web query is first resolved by a domain name server (DNS) that translates the domain name to an IP address. The *domain name* concept is needed to explain why, when a new name is assigned to a server, the name might not be immediately available (since the DNS record has yet to propagate). In another dimension, a richer conceptual view of a web service would recognize that, due to load balancing across servers, an update by one user might not be visible to another user, even after the second user has submitted a later update that was accepted. To understand this, you would need a concept of *web service* that incorporated the notion of eventual consistency.

Understanding the security properties of browsing the web is challenging because so many disparate components and systems are involved, and their functionality is often very complicated. In contrast, application designers seek to present the user with a smaller world—an enclosed garden rather than a busy city—that can be understood in its entirety.

25. **The limits of learnability**. The designer has a responsibility not only to design good concepts, but also to *map* them: to represent them faithfully and compellingly in the user interface to maximize the chance that the user will create a matching mental model.

This is not easy. Some concepts are inherently complex, and some users are less able to infer even simple concepts. As Amy Ko has wisely noted, the standard goal of "learnability" in user interface design should be taken with a grain of salt [82]. Not all concepts can be "taught" through the user interface, and rather than learning through experimentation, most users learn through a social process of engagement with friends and colleagues who demonstrate and explain apps to each other. Most concepts, moreover, are not new but are familiar to users from previous apps, so no single app is solely responsible for teaching a concept.

Of course, apps designed for experts may include concepts whose power is obtained at the cost of some complexity, and in that case, expecting users to be trained is reasonable. The concepts of *channel*, *layer*, and *mask* in Adobe Photoshop, for example, present formidable complexity and confusion to novices, but can be mastered if approached systematically—by learning the concepts one at a time. Learning by trial and error, and watching endless online short video tutorials, is not very effective, despite being the most popular approach.

Chapter 3: How Concepts Help

26. **The power of paragraphs**. As an example of how the *paragraph* concept in Word is stretched to encompass the kind of functionality you'd expect to be provided by additional concepts, consider the formatting of chapters. Rather than having a distinct chapter concept, Word allows you to define a paragraph style—called *chapter heading*, say—using the *style* concept, which, in combination with the *format* concept, allows each paragraph that opens a chapter to be set in an appropriately large

font and given the formatting property *page break before* so that a new chapter always starts on a new page.

Word does in fact have a *section* concept, but it's not what you might imagine. Rather than being a subdivision of a chapter (a concept that does not exist in Word), it represents a contiguous sequence of pages with the same formatting.

27. **Misunderstanding the trash concept.** In one history of the trash, published in Slate, we're told that it was invented "[w]hen the team realized that users needed a way to delete files permanently." But permanent deletion is easy: the hard part was designing a way to get files back! (Cara Giaimo, Why Only Apple Users Can Trash Their Files, *Slate*, April 19, 2016; https://slate.com/human-interest/2016/04/the-history-of-the-apple-trash-icon-in-graphic-design-and-lawsuits.html).

28. **Apple's song concept.** In his biography of Jobs [59], Walter Isaacson noted that, unlike Apple, Sony was not only a leader in consumer electronics but had its own music division—and yet it missed the opportunity to unify those capabilities with the concept of a *song* that could be sold directly to consumers. Even the granularity was new; prior to Apple's innovation, a song could be played but could not be bought: you had to buy a whole album. Although the sleek appearance of the iPod got all the attention when it was first released in 2001, its success can be attributed as much to the *song* concept as to the new technology that made it possible (the FireWire connection that let you upload all your music in an hour, and the microdisk that stored it).

29. **The frequent flyer concept.** The *frequent flyer* concept might even be described as a *dark concept* (by analogy to dark patterns in user interface design) on account of the array of tactics that airlines have developed to make it as unusable as possible. For example, many airlines cause miles to expire but go to great lengths to hide the expiry date from customers, excluding it, for example, in email updates; they make it hard to spend miles by ensuring that very few seats on a given flight are available for reward redemptions; they tempt customers to exchange their miles for low-value magazine subscriptions; and they introduce complex formulas so that discounted seats, and those sold through airline partners, result in fewer "miles." British Airways employed an even more devious tactic: by reducing the apparent cost of a fare and assigning the balance, a much larger portion than for other airlines, to surcharges and taxes, it was able to lower the value of miles (which on redemption covered only the fare itself). A class-action suit over this trick (in a New York federal court in 2018) resulted in a cash settlement of $27m.

30. **Are Gmail labels cost-effective?** According to a study reported by one of the lead designers of Gmail, only 29% of Gmail users had created *any* labels at all [91]. This is not necessarily a problem. It reflects partly the power of Gmail's search mechanism, which obviates the need for manual organization. And it could be that dropping *label* would be a showstopper for some of the three in 10 customers who use it. But it does make one wonder whether the inclusion of *label* is a net win for Google, or whether a different concept might be more cost-effective.

31. **Separation of concerns**. The term was coined by the computer scientist Edsger W. Dijkstra in a 1974 piece [34], in which he quoted from a letter he'd written to a colleague, explaining how separation of concerns entails "study[ing] in depth an aspect of one's subject matter in isolation for the sake of its own consistency, all the time knowing that one is occupying oneself only with one of the aspects."

    "Even if not perfectly possible," Dijkstra explained, separation of concerns "is yet the only available technique for effective ordering of one's thoughts that I know of." Separation of concerns is perhaps not only the most important strategy but also the most underappreciated. It seems so unglamorous, but it's so often at the heart of effective problem solving, and can be the seed for a radical innovation.

    Separation of concerns should not be confused with "divide-and-conquer," a much less powerful technique that comes from the structuring of recursive algorithms. To apply divide-and-conquer, you take a problem and divide it into two or more parts which, if solved, can then be recombined to solve the problem as a whole. Beyond algorithms, divide-and-conquer turns out to have few uses. It assumes that a problem can be neatly divided into parts, and the solutions to the parts easily recomposed—both of which are dodgy premises in most problem-solving contexts.

    A related strategy is *top-down design*, in which a problem is broken into subproblems typically representing processing steps. In the early days of programming, it led to a style of structuring in which processing was broken down into reading the input, doing some computation, and then generating the output. Such a structure is usually poor, being both inflexible (since input and output rarely occur in well-defined phases) and hard to maintain (since the assumptions that readers and writers share are not localized in the code). Separation of concerns, in contrast, would recognize input and output as a subject of interest in its own right, to be treated independently from computation.

    Michael Jackson has noted that top-down development, in which a software system is built by breaking the specification into smaller and smaller parts until eventually the parts are small enough to be implemented directly, usually fails because the initial (and most consequential) decomposition is made when the developer has the least understanding of what's being decomposed [68].

32. **Reusing concept implementations**. Today's libraries and frameworks have the wrong granularity for reusing whole concepts, but efforts are underway to develop new concept-based frameworks that allow programmers to incorporate concepts in their entirety, with both front-end and back-end components, and to configure them flexibly to the problem at hand.

    Déjà Vu is a concept reuse platform built by Santiago Perez De Rosso [119]. Concepts are full stack, including both GUIs and backend services, and are assembled in a variant of HTML, so that minimal programming is required. An end-user developer only needs to pin the existing concepts together (using action synchronization, explained in Chapter 6). In contrast, existing frameworks tend to offer components that are too small (such as the date pickers of GUI libraries) or too large and inflexible (such as rating or commenting plug-ins for content-management systems such as Drupal).

33. **Apple's underpowered synchronization concept**. A cynical observer might suggest that Apple knowingly omits the ability to synchronize only some files in order to encourage users to upgrade their phones to new models with more storage. Apple's designers might also have ruled out selective synchronization because it threatens simplicity. In that case, however, they could have provided a restricted form, rather than supporting the feature in its full generality. For example, they could provide a mode in which only more recent photos are kept on the phone, and older ones are removed while remaining in cloud storage.

    Similar synchronization problems arose with Apple's iTunes. Originally, iTunes was an application whose purpose was to transfer music from a laptop or desktop computer to an iPod music player. The developers of iTunes chose to regard the computer as the holder of master copies of the files, so synchronization was designed to be one way. New files appearing on the computer (bought through the iTunes music store, or ripped from a compact disc) would be transferred to the iPod. But files appearing only on the iPod would not be transferred to the computer, and in fact would be deleted.

    Perhaps this made sense at the time, when an iPod cost less than a tenth of the cost of a computer, and had no means of obtaining files itself except through iTunes. But by 2019, the last year of iTunes' life, the most expensive iPhone had comparable storage and cost to Apple's least expensive laptop, and of course could obtain files through other means.

Confusingly, photos taken on an iPhone camera were propagated back to the computer. But music files were still regarded as owned by the computer and not the phone. The most egregious consequence of this design was that if your computer was damaged or lost, and you purchased a new computer, you would think that iTunes, when synchronizing with your old iPhone would allow you to transfer all your music files to the new computer. In fact, when you connected it, it would warn you that synchronizing would delete all the files on the phone to make it consistent with the new, empty computer!

34. **Two-factor authentication attacks.** The attacks on *two-factor authentication* that I mentioned are two of many that are known. The LinkedIn example is a combination of a social engineering attack (in sending unwary users to a spoof server) and a classic "man-in-the-middle" attack. It was described in 2018 by Kevin Mitnick, a famous hacker who was caught by the FBI in 1995 and spent five years in prison, and later became a successful consultant.

35. **Critical systems: safety and security**. Our society relies increasingly on software for its most vital functions. In the past, "critical systems" occupied a specialized niche, but now almost all software is critical. Few businesses can function at all without working software, and couplings between businesses mean that one failure can have massive repercussions.

Safety and security have traditionally been distinct concerns in software development, applying to different kinds of systems. Safety-critical systems were the ones that posed risks to human life: medical devices, chemical plants, electrical grids and so on. Security-critical systems were those in which damage to data—through corruption or leakage—was the primary risk, and financial loss was the main concern.

This distinction was always tenuous, resting on the faulty premise that we could neatly divide systems into those that controlled physical devices (and were therefore a safety risk) and those that managed our data (and were a security risk). Back in 2005, two anesthesiologists described an incident in which the pharmacy database for a large, tertiary care hospital failed, leaving nurses unable to provide medication for thousands of patients [29]. Catastrophe was averted only because the nurses were able to laboriously reconstruct the pharmacy records from printed slips. And this was long before the era of ransomware attacks on hospitals.

Attention is now turning to security risks in devices. Kevin Fu, a leading security researcher, has founded a center that focuses on security in healthcare, has testified before Congress, and written many papers on the security risks of medical devices, including implantable devices such as pacemakers and defibrillators.

36. **Design flaws in medical devices**. Preventable medical errors account for almost half a million deaths per year in the United States, comparable to the deaths from cancer or heart disease. In a remarkable new book [142], Harold Thimbleby argues that many of these deaths could be avoided with better design of medical devices.

The book is full of shocking examples of user interfaces with egregious flaws at all levels—physical, linguistic and conceptual—and presents a compelling case for better regulation, better training of engineers, and more consistent application of the design principles we already know.

Concepts could play a major role in improving the design of medical devices, by embodying and codifying best practices. It seems that many accidents are associated with just a few areas of practice, and that formulating a shared handbook of concepts such as *dose*, *prescription*, and *adverse interaction* might make it easier to achieve consistent standards across device manufacturers.

37. **Design principles for grounding design criticism**. Principles of user interface design mostly address the physical and linguistic levels. Earlier (in Note 7) I surveyed the development of these principles. In addition to those appearing in Don Norman's *The Design of Everyday Things* [110], explicit lists of principles include Ben Shneiderman's "Eight Golden Rules of Interface Design" [134] (originating in 1985); Jakob Nielsen's "10 Usability Heuristics for User Interface Design" [108] (1994); and Bruce Tognazzini's "First Principles of Interaction Design" [143] (2014). These collections have considerable overlap; the most common themes are the need for consistency, discoverability, visibility, and user control (especially for undoing errors).

Here is an example of some user interface design principles and their application. Don Norman's notion of *affordances* refers to how a physical control (or screen widget) signals to a user (with what Norman calls a "signifier," a word taken from Saussure's theory of semiotics) the capabilities (or "affordances") it offers, and thus the appropriate actions for the user to perform. A door handle, for example, can be shaped to make it clear whether to pull or push, and a user interface button can convey to the user whether to click or slide.

The *Gestalt principles* of visual organization say that items of similar shape and color, and placed close together, will be regarded as offering similar functionality. And Fitts's Law (mentioned in Note 21 and in Figure 2.6) says that the time to move a pointing device to a target is proportional to the distance to the target and inversely proportional to its size.

Together, these principles not only provide a language for critique but may even nudge us towards particular solutions. They might tell us, for example, that an emergency stop button should be large, have a distinct visual appearance, and be close to (but clearly separated from) other buttons.

Similar principles are common in other fields that employ "crits." Typographic designers, for example, talk about the "color" of type, referring not to literal color but to the visual impression of a block of type in terms of its rhythm, texture, and tone. Photographers talk about "balance" in an image (a quality that is hard to define and even harder to achieve, but easy to recognize). And programmers make extensive use of the "minimize coupling" principle, which has many ramifications for avoiding bugs, and for reuse and maintainability (see Note 6).

38. **Design critique vs. user testing**. Design critique is not popular in software design, or indeed in design thinking more generally. In design thinking books and articles, there are endless paeans to brainstorming, but rarely a mention of any kind of analysis or review—which is strange given the prominence of the "crit" in all traditional design fields.

In HCI circles, it seems almost to have become a dogma that user testing is the only appropriate way to evaluate a design. Rather than spending too much time thinking hard about the design (and possibly being misled by prejudices and preconceptions), you are encouraged just to try the design out. You might do this with a prototype, or by building a full product, or by experimenting with variants of the product (a common strategy in web applications).

This kind of evaluation is certainly valuable. Behavioral economics has taught us that human behavior is less governed by reason than we used to imagine, and we are less confident that we can predict how our designs will be received. Software developers often leave user testing too late, maybe even waiting until deployment to discover crippling flaws that will require major revisions of the design. Moreover, the cheapest testing methods—paper prototypes or simple wireframes, and a "Wizard of Oz" setting in which a designer simulates responses to the user's actions—are often the most cost-effective. The excellent online book by Amy Ko [83] gives an informative survey of design methods in general and evaluation techniques in particular.

And with the advent of tools for handling large data sets, and of machine learning as a technique for discovering latent trends in data, empirical evaluation at scale is now cheaper and more effective (although one has to wonder whether the availability of data isn't the cart leading the horse).

Empirical evaluation is also a useful countermeasure when design decisions risk being hijacked by the unsubstantiated opinions of influential people, sometimes referred to as the HPPO (highest paid person's opinions) effect.

The value of user testing was evident in a small app that I built with some students during the pandemic of 2020. Our app aimed to reduce the feeling of isolation during a lockdown by pairing up members of a predefined group, prompting them (by text messaging) to call each other. Early on, we tested a version of the app with some older users. Realizing that the text messages sent by our app might not be identifiable, we had made sure to include in the first message a clear explanation of whom the message was from ("this is the hand2hold service you subscribed to"). But we discovered that these older users, contrary to our expectations, deleted every text message as soon as it was read, so when later messages arrived, they didn't connect them to this initial message, and they dismissed them as spam. The remedy was to include the identification string in every text message.

Despite all this, I can't help thinking that empirical evaluation is overrated. As a means of obtaining feedback on deployed designs, it is obviously necessary. But as a means of creating and refining designs, it may be too crude and expensive a tool. It is infeasible to find a design by exploring a space of options and testing each option empirically. It simply costs too much to implement more than one or two options. At best, we can create a parameterized design, in which some feature can be varied over some range of simple values. These might be the textual content of a message, or the layout of widgets on the page, or the color of some element—all aspects of a design that might turn out to be important in practice, but are hardly fundamental.

The long established approach to evaluating designs involves expert *critique*. The designer—and perhaps others, either as members of a design team or as external reviewers—evaluates the proposed design on the basis of prior experience and deep knowledge of the domain. She anticipates the reaction of users based on the accumulated evidence of how similar users have responded to similar designs in the past. She predicts potential problems by mentally simulating the design, imagining its use in various scenarios, and subjecting the design to known corner cases. And most significantly, the designer—whether consciously or not—applies principles that embody the experience of the community.

Our contemporary fondness for empirical evaluation seems to me to reflect a loss of confidence in the ability of designers to do this effectively. Douglas Bowman, the first visual designer to work at Google, left the company in 2009 in frustration over its approach to data-driven design. In a blog post, he described Google's approach: "Remove all subjectivity and just look at the data. Data in your favor? Ok, launch it. Data shows negative effects? Back to the drawing board. And that data eventually becomes a crutch for every decision, paralyzing the company and preventing it from making any daring design decisions." [14]

Design critique is indispensable, since it is the best way we have to evaluate our work during the process of design. By critiquing ideas as they arise, a skilled designer can navigate a path from a vague intuition to a concrete design proposal, rejecting bad ideas and refining good ones. This kind of work is hard and demanding, and it requires considerable education and experience. This is why, the Pentagram designer Natasha Jen suggests [74], it is downplayed in some design thinking circles, which may be more eager to promote the democratization of design than to recognize how challenging it is.

In summary, then, a constructive approach to design needs both critique and testing, but the balance needs to be redressed, with principled critique forming the centerpiece of the design activity. A designer can apply design principles in the flow of her work, and only when a design satisfies those principles does it make sense to expose it to user testing. This proposed ordering is local rather than global. It does not mean that all design refinement should precede all user testing, but that the design can proceed in small increments—for example, one concept at a time—with user testing after each increment.

39. **An easter egg in Don Norman's book?** Years ago, my brother, Tim Jackson, sharing my delight at discovering Don Norman's book [110], noted a flaw in the design of the book itself which is so self-referential I'm tempted to wonder if it was planted by a designer as an intentional "easter egg."

In the very section explaining Norman's notion of mapping, there is a two-page spread of figures illustrating the application of the idea to the positioning of knobs on a cooktop. In one bad design, for example, the knobs are arranged in a straight line in front of four burners that occupy the four corners of a square; without labels it would be unclear which knob controlled which burner.

There are four figures in the spread laid out not that differently from these four burners. And yet the captions for the figures are placed in sequential order from top to bottom—and, even worse, there are three captions for four figures!

Chapter 4: Concept Structure

40. **Purposes aren't goals.** You might wonder: why isn't *deletion* a perfectly reasonable "purpose" for the *trash* concept? After all, isn't that what people use it for most of the time? (See Note 27.)

The power of purpose as a design notion would be weakened, however, if this were admitted. Instead, I prefer to make the following distinction. A *goal* is something that a user may want to attain at one time or another, and which may be achieved with the help of a concept; a *purpose*, on the other hand, is a need that motivates a designer to include a particular concept. So goals for the *trash* concept might include deleting files, tracking information about previously deleted files, and even providing soothing ASMR (autonomous sensory meridian response) sounds by repeated emptying. Undoing deletion would be a goal too, but out of all these, it's the only purpose, because there are simpler (and in some cases more effective) ways to provide the other goals.

Or to put it another way, if the *trash* concept were not there, what opportunities would be missing? On the reasonable assumption that there would some way to delete things (since every prior operating system that could create files had an easy way to remove them), this suggests that the purpose of the *trash* is undoing deletion, since that's the key benefit that it brings.

Adding a concept to a design generally increases the cost, both to the designers and developers who build the software, and to the customer who must learn how to use it. The concept's purpose helps you decide whether the cost is justified by the benefit, by articulating the precise advantage that this concept brings and thus the motivation for including it.

The same distinction between purposes and goals can be applied to the other two examples in this chapter. The purpose of the *style* concept is to enable consistent formatting of documents—in particular, allowing you to modify the format of a single style and have all the elements of that style updated in concert. Naive users of programs like Word often think that styles are for *applying* formatting: that they just give you a quick way to apply a collection of predefined formatting properties to an element (for example, making a section header bold and large). But as with the deletion goal of the *trash* concept, there are much simpler ways of achieving this goal without the *style* concept.

And likewise for the *reservation* concept: a user's goal may be simply to obtain a table at a restaurant. But if that were all the restaurant wanted to provide, and it was happy to have customers turning up and waiting for tables, it could use a simpler concept in which tables are just assigned on the fly.

The trash is not a metaphor. Another common misconception about the trash is that it's attractive and intuitive to users because it's a *metaphor* for a physical trash can. Metaphors do play a role in many user interfaces: folders and files, for example, clearly benefit from drawing an analogy to physical folders and the paper files they contain.

For understanding the *trash* concept, however, a physical trash can is a misleading metaphor. The purpose of a physical trash can is to *stage* the removal of trash: we have trash cans in our kitchens so we don't need to take each item of garbage outside to the larger bins that are collected weekly. The purpose of Apple's *trash* concept is, instead, to allow the *restoration* of items from the trash—a feature also supported by the kitchen trash but not one we generally like to take advantage of.

41. **The importance of names in design.** Names are essential in design, because they allow us to refer to known patterns without having to explain them. Just as an architect might say "let's have a cantilevered balcony" rather than "let's add a little platform on the side of the building that people can walk out onto, and which doesn't have visible supports holding it up," so a software designer might say "let's use the trash concept for handling message deletion" rather than "let's handle deletion of messages the way they do on the Mac desktop with that trash can icon that lets you restore files that you deleted."

42. **State is memory.** Another way to describe the state of a concept is that it captures what the concept must *remember* during execution. So if the items in the trash were to be shown sorted by time of deletion, we would need to remember when each item was deleted, and the association between items and their times would have to be included in the concept's state.

I find that this simple idea is often not clear to novices constructing data models. They somehow come to think of a data model as if it were making truth statements about the world. This view is encouraged by using "has" terminology for relations (as in "a user has a name"), and it's not implausible when designing pure ontologies (for example, a data model intended to describe chemical compounds). But it's rarely a helpful perspective in software design, where the data model is there only to support actions, so if no action requires a particular piece of information, it can be omitted from the data model, and conversely, if an action relies on some information, it must be included.

Concepts make data modeling even more straightforward, because in place of a global data model, there are instead a collection of miniature data models, each defined as the state of a single concept, and justified by that concept's actions.

43. **The operational principle.** The operational principle can also be viewed as a *theorem* about the behavior of the concept, asserting that the concept has some key property, or equivalently as a kind of generalized test case. But whereas theorems and test cases play a secondary role, being used to check the consistency of a program or specification, the operational principle is front and center in the design of

the concept: it's the very essence of the concept (and often the most compelling way to explain it).

Imagine Isaac Newton—presumably a bright fellow—examining a toaster, and wondering what it does and what it's for. He sees a knob that's labeled as going from "light" to "dark," a lever you can press down, and a button marked "cancel." How would you explain the toaster to him? You surely wouldn't start by describing the state (the dial setting that determines the cook time, the on/off state of the heating element, and the state of the timer, with the time remaining until ejection). Nor would you start with the actions: for example, that turning the dial adjusts the state component corresponding to the cook time.

You'd just tell him what the toaster is *for*, namely its purpose, which is toasting bread. Then you'd demonstrate *how* to use it—its operational principle—by putting two slices of bread in the top, setting the knob halfway, say, and pushing the lever down. You might extend the operational principle to a second scenario by setting the knob too high, and pressing the cancel button to eject the toast when it begins to burn.

So although the state and the actions define the behavior fully, they don't provide the crucial missing parts that both designers and users need: the need for the concept (given by the purpose) and the demonstration of how that need is fulfilled (given by the operational principle).

The operational principle comes to me via Michael Jackson [72] from the work of the philosopher Michael Polanyi [125] who explains that the operational principle of a machine specifies how "its characteristic parts—its organs—fulfill their special function in combining to an overall operation which achieves the purpose of the machine."

My toaster example is based on Polanyi's observation that the laws of physics are incapable of explaining a machine—a fundamental distinction between physics and engineering, for which a "logic of contriving" is required. In Polanyi's words: "Engineering and physics are two different sciences. Engineering includes the operational principles of machines and some knowledge of physics bearing on those principles. Physics and chemistry, on the other hand, include no knowledge of the operational principles of machines. Hence a complete physical and chemical topography of an object would not tell us whether it is a machine, and if so, how it works, and for what purpose. Physical and chemical investigations of a machine are meaningless, unless undertaken with a bearing on the previously established operational principles of the machine." [124, p. 39]

Thimbleby's partial theorems. Harold Thimbleby has noted that users infer behavioral theorems by observing their interactions with systems, and come to rely on them [140]. This is fine if theorems are either true (in which case they will often be helpful) or manifestly false (in which case users will quickly recognize their invalidity).

Trouble comes with what Thimbleby calls "partial theorems": theorems that are *almost* always true. Because a partial theorem holds most of the time, the user will come to believe that it holds universally, but one day will be misled and get a nasty surprise. For example, in an email application with folders, it is natural to assume that the "sent messages" folder (1) contains every message that was sent and (2) contains only messages that were sent. Both properties are in fact untrue: the first (more obviously) because messages can be moved into other folders to organize them, and the second because you can (for no good reason) move a message that you received but did not send into that folder.

Here are some examples of partial theorems for the three concept examples of this chapter. Each of these theorems is true for the concept itself, but partial in some implementations. For the *trash* concept, the operational principle as I've stated it is actually a partial theorem for the Apple Macintosh implementation. As I explain later in Note 73, emptying the trash does not always permanently remove every file that is in the trash (although most users will never observe this case).

For the *style* concept, there is a common extension in which styles can be arranged in a hierarchy, with each style inheriting the formatting properties of its parent. In one of my first papers [62] on automatic analysis of software models, I used the *style* concept as my running example, and showed how a rather mundane theorem turns out to be partial. You might think that if you change the parent of a style, changing it back immediately will have the effect of an undo, and the styles will be unchanged. Surprisingly, this turned out not to be the case for common implementations. I won't spoil the surprise by telling you why in detail, but will just hint that it arises when a style and its parent set the same value for a given formatting property.

For the *reservation* concept, the obvious partial theorem is again the operational principle itself—and its partiality is understood by most restaurant customers. When you turn up at a restaurant with a reservation, there is a small chance that no table will actually be available. The reason is that the restaurant, in accepting reservations, must make an assumption of the maximum length of time that a party will remain at a table. Sometimes a party will overstay its welcome, and the restaurant may find it lacks enough empty tables to honor reservations.

The operational principle vs. use cases and user stories. Scenarios are popular in many approaches to software development. Ivar Jacobson [73] developed an approach to requirements that involved specifying "use cases," each a scenario in which the user interacts with the system and achieves some goal. "User stories" are similar but are usually expressed in a less structured way.

Both of these are unlike the operational principle. They play a different role. Rather than explaining the essence of the design, they are intended to describe the functionality in its entirety. Moreover, use cases and user stories are applied at the system level, and not at the more granular level of individual concepts.

So a full requirements description might have dozens or hundreds of use cases, handling not only normal usage but abnormal cases too. In concept design, you give only the few scenarios that are needed to show how the concept fulfills its purpose, and the details (including the abnormal cases) are handled by the action specifications. In the *style* concept, for example, the operational principle need not include a scenario explaining what happens when a style is deleted. This is a tricky case, but it's hardly the essence of the design.

The essential quality of the operational principle, distinguishing it from regular scenarios, is the way in which it motivates the concept's design as a whole. Suppose you asked me how *style* worked, and instead I showed you that you can define some styles, such as *body*, *quote* and *heading*, and then format a paragraph easily by selecting the relevant style and applying it. This is a perfectly valid scenario, and people do it all the time. But it's not the *motivating* scenario for the *style* concept. To support this simpler scenario, the app doesn't need to remember which styles are associated with which paragraphs. A less powerful concept would suffice.

Use cases and user stories are commonly used as increments of programming work in agile development. This is an appealing idea, and consistent with the agile goal of incrementally delivering pieces of functionality with clear value to users. But focusing on just one or a handful of use cases may give too incomplete a picture, leading to an implementation that can't easily be extended to the next use case.

Suppose, for example, you were implementing the *style* concept, and you started with use cases in which you define a style and apply it to some paragraphs. If these use cases don't involve *modifying* an existing style, your implementation wouldn't need a mapping from paragraphs to styles. When you apply a style to a paragraph, you could just take the format settings defined in the style and apply them directly to the paragraph. This is a plausible concept—just a different one. It goes by the name of "style" in Apple's TextEdit app, but by omitting the crucial association of

styles to paragraphs, it fails to be very useful. Only later when you get to implementing a use case in which a style is modified would you discover that your code
cannot be easily extended. In fact, including in the use case a style that is modified
won't be enough; it will have to be a style that is modified *after* it has already been
applied to a paragraph.

Of course this is a small example, and reworking the code might not be so hard
in this case. But in the context of a large system, these kind of problems can become crippling, and suggest a reason for why teams that follow an agile approach
often program themselves into a corner and have to throw work away (a practice
also known as refactoring).

Concepts offer a better granularity for incremental development because they
are self-contained and largely independent of one another. If you were implementing a word processor with concepts, you might start with the concept of *paragraph*
and build the basic code for editing text (and dividing it into paragraphs); then you
might implement the *format* concept that lets you change the size and typeface of
the text, for example; and then you might implement the *style* concept that allows
you to apply formats systematically to paragraphs through their styles.

44. **Concept formalism**. The concept descriptions in this book are given informally, in plain English, in order not to put off readers unfamiliar with code or logic. In
practice, it's often better to use a formal notation that is precise and unambiguous,
and can be potentially be analyzed automatically and compiled into running code.
Many notations exist that would be suitable for specifying concepts; these include
Alloy [66], Z [136], VDM [75] and B [1]. Alloy does make a cameo appearance in
the concept descriptions in the declarations of the state components, and for the
constraint (in Figure 4.4) that defines *format* in terms of the other components.

Concept semantics. For the benefit of computer scientists and other readers wanting a more rigorous explanation of what a concept definition exactly defines, I am
including here some notes on concept semantics. I'll focus on concept *behavior*;
the purpose of a concept does not seem to be readily formalizable.

A concept's behavior can be viewed, in the simplest terms, as a state machine.
Figure E.3 shows part of the state machine for the *trash* concept (Figure 4.2). In the
initial state, no items are accessible or trashed, and the only possible action is to
create an item. Executing the action *create(i0)* leads to a new state, in which the item
i0 is now accessible; *delete(i0)* takes us to a third state in which the item is trashed,
and there are now two options: *restore(i0)*, which undoes the deletion, and *empty()*,
which returns us to the initial state.

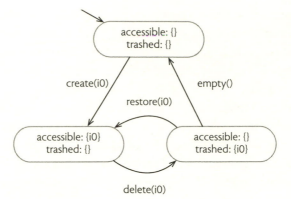

FIG. E.3 *Part of the state machine for the trash concept.*

This little state machine only includes actions on a single item, *i0*. The concept itself allows *Item*, the set of possible items, to be unbounded, so the real machine is infinite. Nevertheless, even if we can't draw it, we can define it. You start with the initial state, and then consider each action in turn. If the action definition says it's allowed in that state, you add a transition to the state it prescribes, repeating this for each possible action. You then find the outgoing actions for any new states that were produced, and continue in this fashion, expanding the machine for ever.

(A full concept definition language would include some syntax for defining the initial state. I haven't needed this in any of the examples, because the states have always comprised sets and relations which are assumed to start out empty.)

Transition relations, preconditions and deadlock. Mathematically, the meaning of an action is a *relation*. Given a concept with a set of states *S*, an action *A* with an argument drawn from the set *X* has an associated relation

$trans(A) \subseteq S \times X \times S$

containing all the transitions *(s, x, s′)* of the action, where *s* is the state before, *x* is the argument, and *s′* is the state after. To account for more than one argument, we just view *X* as a set of tuples.

The transition relation need not be total. The set of pairs *s, x* of prestates and arguments for which a poststate *s′* exists is called the *precondition* of the action. When the precondition does not hold, the action is not enabled and may not occur. (This is subtly different from the notion of a precondition of a function in programming. There, with nothing to prevent a function from being called, the precondition represents only an obligation: if you call a function when it does not hold, the usual assumption is that anything can happen.)

In any given state, any number of actions may be enabled. It's possible to design a concept in which a state can be reached from which *no* action is enabled. This is called *deadlock*, and is undesirable because it means that the concept can never do anything else again.

Usually a concept embodies a collection of objects, each with its own lifetime. The lifetime of any individual object may well come to an end, but the behavior of the concept as a whole continues because the collection itself is not bounded. The *reservation* concept, for example, handles the lifetimes of individual resources—a dinnertime slot at a restaurant, for example—which come to an end when the resource is used. But a new resource can always be added, so the concept as a whole continues to execute, and there is no deadlock.

Formalizing actions. States and actions can be readily formalized in any of the languages mentioned above. For example, the *reserve* action that was defined informally in Figure 4.7 as:

```
reserve (u: User, r: Resource)
  when r in available
  associate u with r in reservations and remove r from available
```

can be written formally in Alloy like this:

```
pred reserve (u: User, r: Resource) {
  r in available
  reservations' = reservations + u -> r
  available' = available - r
  }
```

The precondition is the same (because *in* is a keyword in Alloy already), but does not need the keyword (because preconditions are implicit in Alloy). The last two lines define the new values of the two state components (denoted by priming their names) in terms of their old values. (The priming, by the way, is a shorthand introduced to Alloy by the Electrum [20] extension.)

If I were designing a new language for concepts (which I hope to do), I would prefer a more operational syntax in which relations can be updated on particular domain elements, writing something more like this:

```
reserve (u: User, r: Resource)
  when r in available
  u.reservations += r
  available -= r
```

with exactly the same intended meaning. Following the programming language CLU, this uses C-style shorthands in a generalized way, so a statement of the form *e1 op= e2* stands for *e1' = e1 op e2*. The second line above is thus short for an update constraint along with an implicit frame condition:

u.reservations' = u.reservations + r
all p: User | p != u implies p.reservations' = p.reservations

Traces and state observations. The state machine formulation has a simple and mechanistic quality to it, but in practice a more abstract view is helpful. We can define the *traces* of the concept to be the set of all the finite histories of action instances that are possible. So for the *trash* concept, these include:

<>
<create(i0)>
<create(i0), delete(i0)>
<create(i0), delete(i0), empty()>
<create(i0), delete(i0), restore(i0)>
...

and so on. In addition to observing these traces, the user of a concept can observe the state at any point. That is, there's some function

state: Trace → State

that maps each trace to the state that it produces. The state produced by the empty trace, *state(<>)*, is of course just the initial state. The state after a longer trace can be computed by applying each action in turn, following the rules given in the action definitions. So after deleting an item we find it in the trash:

state (<create(i0), delete(i0)>) = {accessible: {}, trashed: {i0}}

and after emptying the trash, the item is permanently removed (and coincidentally we're back in the initial state):

state (<create(i0), delete(i0), empty()>) = {accessible: {}, trashed: {}}

System actions and determinism. Concepts are assumed to be *deterministic*. This means that the transition relation associated with each action is functional: given a particular state and set of argument values for which an action is enabled, at most *one* state can result from executing it.

This is what justifies the definition of the state function: a trace always leads to exactly one state. Another consequence of determinism is that an action cannot be refused arbitrarily; whether it can occur is defined (by its precondition) in terms of the state in which it is invoked, so if you know the trace that has occurred up to some point in time, you can predict whether a given action can follow.

Not all actions need be performed by the user. In general, different categories of users will perform different subsets of actions. System actions, performed spontaneously without any user participation are possible too, and in a full concept notation might be marked with a special keyword. For example, in a flight reservation system, the *seat allocation* concept may have an action

system assign-seat (c: Customer, s: Seat, f: Flight)

that assigns to customer *c* the seat *s* on flight *f*. Since the system (and not the user) selects the seat, this allows us to describe the scenario in which a user, having bought a ticket, gets a seat assigned, apparently arbitrarily. In some sense this is a non-deterministic outcome of the user's request, but because the choice of seat is visible in the argument of the action that the system takes, the concept remains deterministic. And indeed, in practice—in a good design at least, even if not the design that many airlines choose—such an action would be synchronized with a notification letting the user know which seat had been assigned.

A logic of operational principles. Operational principles can also be readily formalized. Different kinds of temporal logic might be used; here I will illustrate how they might be expressed in *dynamic logic* [53] because it most closely matches the style of the informal definitions that I have shown.

The basic forms of dynamic logic are *[a]p*, which says that after performing the action *a*, the predicate *p* always holds, and *<a>p*, which says that after performing the action *a*, the predicate *p* may hold. For the operational principles we'll write, only the first form is needed, and I'll drop the modal operator, writing *a{p}* instead of *[a]p*.

Actions can be combined into compound actions using *sequential composition*, with *a;b* denoting the *a* followed by *b*; *repetition*, with *a** denoting zero or more occurrences of *a*; *choice*, with *a or b* denoting an occurrence of either *a* or *b*; and *negation*, with *not a* denoting any action that is not the action *a*. I'll also assume a special operator *can* that extracts the precondition of an action; thus *can a* holds in a state in which the action *a* can happen.

With these operators in hand, we can formalize the operational principles of this chapter. For the *trash* concept, the principle written informally as

after delete(x), can restore(x) and then x in accessible

becomes two separate formal assertions:

delete(x) {can restore(x)}
delete(x); restore(x) {x in accessible}

For the *style* concept, the informal principle

after define(s, f), assign (e1, s), assign (e2, s) and define (s, f'), e1 and e2 have format f'

becomes

define(s, f); assign (e1, s); assign (e2, s); define (s, f') {e1.format = e2.format = f'}

And for the *reservation* concept, the informal principle

after reserve(u, r) and not cancel(u,r), can use(u, r)

becomes

reserve(u, r); (not cancel(u, r)) {can use(u, r)}*

There's some latitude in exactly how the operational principle is expressed, and formalizing it brings the various options to mind. For example, you might wonder why, given that the principle for *reservation* allowed for actions between reserving and using a resource (while stipulating that no cancellation occur), the same flexibility wasn't expressed in the *trash* concept's principle. In the latter case, we might have written instead

delete(x); (not (restore(x) or empty())) {can restore(x)}*

saying that after deleting an item, you can perform any other actions prior to restoring it except for restoring it (since you can't restore it twice) and emptying the trash.

My (admittedly flimsy) justification for not expressing it this way was that the scenario expressed by my original principle for the *trash* is not rare: it's common to delete an item by mistake and restore it immediately. But for the *reservation* concept, a scenario in which a reservation leads to a use without any intervening reservations occurring is impractical and thus a poor exemplar.

A linear-temporal logic version. The operational principle might also be expressed in *linear temporal logic* (LTL) [123]. LTL and dynamic logic are not equivalent in their expressiveness, but both can express the general form of our operational principles—namely, that after some possible sequences of actions, some condition holds. In the newest version of Alloy [20], which includes LTL operators, the operational principle for the *reservation* concept

after reserve(u, r) and not cancel(u,r), can use(u, r)

can be written

all u : User, r : Resource |
  always reserve[u, r].then [can_use[u, r].while [not cancel[u, r]]]
  }

having defined the *then* and *while* operators as macros:

let then [a,b] {a implies after b}
let while [a,b] {not b releases a}

Although this formulation seems to me a bit less intuitive than the one in dynamic logic, it has the great advantage of being usable right now in Alloy, and can be executed and checked automatically. (Thank you to Alcino Cunha for providing this example.)

Real actions in temporal logic. In fact, neither dynamic logic nor linear temporal logic exactly captures my intended meaning for operational principles, because both treat the occurrence of an action as equivalent to its state transition.

In the operational principle of the *reservation* concept (Figure 4.7), for example, the expression *cancel(u,r)* matches—in both logics—*any* transition resulting from an execution of the *cancel* action with arguments *u* and *r*. If the *retract* action were defined (differently) to allow a resource to be retracted even if reserved, then, after the action *reserve(u,r)*, the actions *retract(r)* and *cancel(u,r)* would have exactly the same effect (namely undoing the reservation of resource *r* by user *u*). And yet we might want the mention of *cancel(u,r)* in the operational principle to apply only in a case in which a reservation by user *u* of resource *r* is dropped due to a cancellation and *not* a retraction. To make this distinction, a richer semantics is required in which the *names* of actions are significant.

Incidentally, although its title might suggest otherwise, the Temporal Logic of Actions [85], an elegant logic devised by Leslie Lamport and supported by a powerful model checker, also doesn't represent actions explicitly, and, like Alloy's temporal logic, might be more accurately (if less catchily) described as a "logic of state transitions."

Classification of objects. I've talked about an action instance such as *create(i0)* happening, but I haven't explained what kind of thing the object *i0* is. Because identifying and organizing objects is such an important part of the design process, it's useful to classify objects into different kinds. I'll describe three different dimensions along which objects can be classified: by role, by mutability, and by interpretability.

Object roles. Objects play three different roles in a software system. First, an object may play the role of an *asset*. An asset has inherent value, although the value may be different for different users at different times and for different purposes. Some assets—photos, audio tracks, blog posts, comments—correspond to familiar objects in the physical world; others, especially those associated with security—certificates, permissions, capabilities, passwords, etc.—are more abstract.

Second, an object may play the role of a *name*. Names are used to locate or identify other objects. Sometimes these named objects are physical: a social security number names a person; a serial number names a camera; a mailing address names a building. Sometimes they exist in the world but are not physical: a date names a day in the past or the future, and a product code names a product category (corresponding, for example, to all Roxbury russet apples). And sometimes they are virtual, and exist only inside a computer: an email address names an email account, a domain name names a server, a file system path names a file or folder.

A name is usually intended to be unambiguous, pointing to exactly one object, but its interpretation will often depend on context. A small company, for example, may name its members using first name/last name combinations (until the day a second John Smith comes along). A name for one object may be used as a proxy name for another; it's common, for example, to use phone numbers as names for people, since almost everyone now has a unique mobile phone number.

The third role of an object is as a simple *value*. Values, unlike assets and names, carry no meaning of their own; their meaning comes from their relationship to other objects. The number 80, for example, may be a person's age, or the temperature of water in a lake, or the number of hits a website has received in the last hour.

Figuring out what role an object plays is an essential part of design, and prompts some immediate design questions. For an asset: Who owns it? Is privacy a concern? How would you search for it? For a name: What context is it interpreted in? Is it indeed unique? How long is it valid for? For a value: What does it describe? Does it have units? Can two different values be usefully compared?

The same object can play different roles in different concepts. On the Hacker News website, users upvote and comment on posts that comprise no more than a link to a web page. For the *post* concept itself (and most other concepts of the site), a post is an *asset*; but to the *url* concept that provides the functionality of following links, a post is a *name*.

Object mutability. A mutable object is one that changes over time. For such a notion to make sense, we need to distinguish the *identity* of the object from its *value*. (I'm using "value" here as it's conventionally used in programming, and not with the specialized meaning of "simple value" in the role classification just discussed.) To "change" means that the value associated with a given identity at one point in time differs from the value associated with the same identity at another point in time.

Where exactly the identities lie, and which values are replaced, is usually a subtle question, and is determined more by choices of description than by any objective reality. Take, for example, an image in Photoshop that is comprised of some two-dimensional array of pixels. Suppose we apply an adjustment to the image that makes it appear darker. Did we change the value of the image itself, so that now it contains a different array of pixels? Did we keep the array but change which pixels it contains? Or did we keep both the array and its pixels, and change the values of the pixels themselves? An examination of the code might reveal which of these is correct from the perspective of the programmer. But for the user, it's hard to argue that one is more compelling than the other.

Identities, by the way, are not the same as names. The relationship between an object's identity and its value is more like that between a box and its contents. A name allows you to *refer* to an object, and often also to *find* it. An identity, in contrast, merely allows you to *distinguish* one object from another.

Since concepts communicate only by the synchronization of actions (as will be explained in Chapter 6), objects that are passed as arguments of actions are required to be immutable. If they were not, an action in one concept might mutate an object shared by another, leading to a hidden communication. This would open up the kinds of complications (such as aliasing) that plague programming languages and have led language designers to control if not eliminate mutability whenever possible.

Within a concept, the user is free to interpret the concept state in terms of mutable objects. Such an interpretation is no more valid, however, than one in which all objects are immutable and changes involve only the relations between them. In our description of the *reservation* concept (Figure 4.7), for example, there is no explicit "reservation" object; instead, a reservation is represented as a tuple in the relation *reservations* from users to resources.

Interpreted and uninterpreted types. The third and final dimension along which objects can be classified is whether they are *interpreted* or not. Most of the objects involved in a concept are treated as if they were opaque. The concept behavior only recognizes equality: that is, whether an object held in one variable or state component is the same object as that held in another.

For example, the *cancel(u,r)* action of the *reservation* concept (Figure 4.7), which cancels user *u*'s reservation of resource *r* compares the objects named *u* and *r* with the objects it stores in the *reservations* relation. From the action's point of view, users and resources could be anything so long as one user or resource can be distinguished from another.

In contrast, consider a *rating* concept with an action *rate(i,n)* in which a user rates an item *i* with a number *n* (between 0 and 5, say). In this case, the item is uninterpreted, but the number *n* is interpreted as an integer, and contributes to the computation of the item's average rating.

Why does this distinction matter? If the same type flows from one concept to another—by synchronization, as explained in Chapter 6—the two concepts will need to treat the objects of that type in a consistent way. For uninterpreted types, it's trivial; they must simply agree on which references to an object denote the same object and which do not. But for interpreted types, it's more challenging, because the two concepts must give the same interpretation to each object. If the shared object is a date in the form 050721, for example, the concepts would have to agree on whether this represents May 7 (following the typical American order of mm/dd) or July 5 (following the British order of dd/mm).

The distinction between interpreted and uninterpreted types can be formalized in terms of *permutation invariance*. Each type T that appears in a concept description has some set $Objs(T)$ of objects associated with it, comprising the objects that variables of that type can hold. The permutations of T are all the one-to-one functions from the set $Objs(T)$ to itself.

Applying a permutation p to an action instance a $(x^0, x^1, ...)$ in which argument x^0 has type T^0, x^1 has type T^1, etc., gives the instance a $(y^0, y^1, ...)$ in which y^i is $p(x^i)$ if $T^i = T$ and x^i otherwise. Permutations are lifted over traces and states in the obvious way.

Now can we express the invariance property: a type T is uninterpreted in a concept C if, for every permutation p of T, and for every trace t of C, the permuted history $p(t)$ is also a trace of C, and $state(p(t)) = p(state(t))$.

For example, given this trace of the *reservation* concept and its associated end state

    <provide(r0), reserve(u0, r0), cancel(u0, r0)>
    available: {r0}, reservations: {}

applying the permutation that swaps resources *r0* and *r1* gives

    <provide(r1), reserve(u0, r1), cancel(u0, r1)>
    available: {r1}, reservations: {}

which is also a valid trace and end state. Since this invariance condition would hold for any permutation of *User* or *Resource* and for any trace, both types are uninterpreted.

This is hardly surprising, since these types are generic type variables, so by definition the concept can assume nothing about their objects. But a regular, non-generic type can be uninterpreted too. The *Task* type in the *todo* concept (Figure 6.2) is uninterpreted, for example, because the concept includes no functionality such as searching for tasks with given content, or checking spelling, which would require looking "inside" a task.

To check whether a type is interpreted, one need only look at the operations performed on its objects in the state updates of the actions: if they include only the "logical" operations of sets and relations and equality comparisons, the type is uninterpreted. (Alfred Tarski, who invented the relational calculus that forms the basis of Alloy, characterized "logical notions" as being exactly those whose meaning is invariant over permutations of the universe [138].)

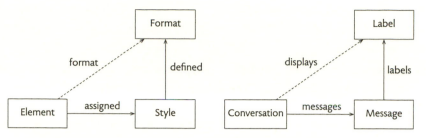

FIG. E.4 *Data models for the style concept (left) and labels in Gmail (right).*

45. **The power of relations in data modeling.** In defining a derived state component in the *style* concept (Figure 4.4), I wrote

   format = assigned.defined

which says that the relation *format* is the relational composition or "join" of the relations *assigned* and *defined*. Such constraints can be given a nice diagrammatic interpretation (Figure E.4, left): the path comprising the arrow labeled *format* is equivalent to the path comprising the arrows labeled *assigned* and then *defined*.

   Or in words, more laboriously: to get the *Format* associated with an *Element*, you can either follow the *format* relation in one step, or you can get the associated *Style* by following the *assigned* relation in a first step (getting the style assigned to that element), and then follow the *defined* relation in a second step (getting the format defined by that style). The diagram also makes clear, incidentally, how the design exploits indirection: see *Concepts have rich states* in Note 48.

   Thinking of relations as arrows and paths in a graph is intuitive and helpful when you get the hang of it, and is reason alone to describe your data model in terms of relations (and to draw it as a diagram). Just looking at a data model diagram and considering how paths might be related is an easy way to discover possible constraints. Figure E.4 (right) shows a similar situation for labels in Gmail, in which a constraint of the same form applies (and turns out to be the source of some serious problems, as explained in Chapter 8).

   This style of modeling is the essence of Alloy [66], and was inspired by Z [136], which pioneered it. Most other notations do not support this kind of navigational thinking. In first-order logic, you'd need quantifiers:

   $\forall$ *e: Element, f: Format | (format(e, f)* $\Leftrightarrow$ $\exists$ *s: Style | style(e, s)* $\wedge$ *defined(s, f))*

In an object-oriented style, you might think you could write something like this:

   $\forall$ *e: Element | e.format = e.assigned.defined*

but it would be ill-formed because of the cases in which *e.assigned* is undefined. A solution to this problem is to treat such an expression as denoting a *set* of values

235

(which is empty when an element is not mapped), and to lift navigations over sets. This is exactly what Alloy does, by treating the mappings as relations and the dot as relational join.

46. **Near misses aren't necessarily bad.** Calling a concept a near miss doesn't mean that it's necessarily a bad design. The *style* concept brings some real complexity in the structure that must be maintained. Apple's color picker offers a system-wide palette, so a document in a particular application would need somehow to name this external palette, introducing a dependence that has its own liabilities.

47. **Bad conflict detection: an airline reservation example.** A common feature in variants of the *reservation* concept is *conflict detection*, in which users are prevented from making reservations that inherently conflict, in order to prevent them from gaming the system and making reservations they're likely to cancel.

    Sometimes the conflict rules get in the way of reasonable usage. My wife once wanted to fly from Boston to Cincinnati on a Sunday morning for a wedding that afternoon, and to return home that night. A round-trip ticket wasn't available, but there were open flights in both directions. So she tried to make two reservations, one for the outgoing flight and one for the return flight. The airline's system wouldn't allow her to reserve both flights at once, and displayed a message saying that you couldn't take flights from two different cities on the same day!

48. **Concept characteristics.** Because the main text only hints at some of the essential qualities and characteristics of concepts, I thought it might be helpful to elaborate these for readers wanting more detail.

    Concepts are inventive. When philosophers talk about concepts, they're usually referring to the way we categorize things that already exist in the world. Thus we use concepts like "dog" and "cat" to classify animals. In the early 18th century, Carl Linnaeus developed the taxonomy we still use today for classifying biological organisms, by defining concepts based on characteristics. Birds, for example, lay eggs, have feathers, have warm, dark blood, and "fly in the air and sing."

    (Linnaeus's contribution was more the idea of classification than the classification itself. The *Ornithological Dictionary; Or, Alphabetical Synopsis of British Birds* of 1831 complained that his classification wasn't entirely accurate, noting that "more than two-thirds of the known species of birds *never sing*; and that many birds, such as the cassowary, the ostrich, and the penguin, *cannot fly* for want of sufficient wings.")

    In contrast, when we talk about the concepts of a software app or service, we're referring to things that were themselves *invented*. They might not have been invented by the designers of the app, but they were invented by somebody at some time. Your tax preparation app didn't originate the concept of *social security number*. That concept was not just "out there" in the world along with dogs and cats, however. It was invented, in 1936, to keep track of the earnings of American workers for administering their benefits under the new Social Security program.

The *trash* concept was invented by Apple, and was a focus of its legal struggles with Microsoft and Hewlett Packard. In the end, Apple lost its claim to ownership of the concept but retained the right to the name (which was why it was called the "recycle bin" in Windows).

The *style* concept was invented by Larry Tesler and Tim Mott at Xerox PARC in the early 1970s. Styles were introduced in Bravo, the first WYSIWYG document preparation system, developed at Xerox PARC in the early 1970s by a team led by Butler Lampson and Charles Simonyi. PARC's Larry Tesler and Tim Mott had visited Ginn, a traditional printer in Boston, to learn about how text was marked up, and conceived the idea of styles, which Simonyi subsequently included in Bravo, and then took to Microsoft, where he led the development of Word.

The *reservation* concept appears to have been invented, at least for restaurants, in the 19th century, when big city restaurants began to offer advance bookings for both tables and private rooms. It's a fun game identifying concepts that are essential pieces of our social infrastructure and uncovering their origins. Alexis Madrigal, a writer for *The Atlantic*, researched the restaurant reservation, with help from historian Rebecca Spang. Why reservations took off only at the turn of the 20th century isn't clear, but seems to be related more to social changes that made restaurant dining more pervasive than to the advent of the telephone. (See: Alexis Madrigal, "Where restaurant reservations come from: A journey into the mysterious origins of the pre-arranged table." *The Atlantic*, July 23, 2014.)

Concepts evolve. As the benefits and limitations of concepts become clear over time, their designs evolve. Small improvements may accrue over years or even decades. The *style* concept was quite primitive when it first appeared in the Bravo text editor at Xerox PARC. There was a fixed collection of built-in styles, and although users could adjust the formatting properties of a style, they could not create new styles. (This is actually the situation in the Google Docs app today.) The same restriction carried over to the first version of Microsoft Word, but eventually the concept expanded to include user-defined styles, as well as stylesheets for packaging collections of styles.

A nice example of a design improvement that took many years to come can be found in the *group* concept. Used in all drawing and layout programs, it allows the user to bind multiple objects together so they can subsequently be treated (in many respects) as a single object. For many years, however, the design had an annoying flaw. If you grouped some objects together, and later wanted to make a slight adjustment to one of the constituent objects, you had no choice but to ungroup the objects, select and modify the object in question, and then regroup them. This was especially burdensome if there were many objects in the group, and if the group object itself had its own assigned properties (such as animation order) that were lost when ungrouped.

About ten years ago, decades after the introduction of the *group* concept, Apple designed a clever solution that has since become universal. A single click will select an object, and if the object belongs to a group, the entire group will be selected as a single object. A double click will select an object *within* a group, which can then be individually edited, essentially "opening" the group. A click on another object within the group will select that object; a click on an object outside the group, or on the background, will unselect the grouped object and "close" the group, leaving it in its previous state all but for the modifications made to the constituent objects.

Some concepts evolve in a different way, with their behavior responding to a changing purpose. Consider Facebook's *post* concept, for example. Initially, posts were limited to 160 characters. The limit was progressively extended, to 420 characters in March 2008; to 5,000 in September 2011; and then beyond 60,000 in November 2011, only two months later. The *post* concept thus morphed from providing a short message to supporting publishing of long articles—a very different purpose. Indeed, in advertising the latest limits, Facebook noted that a typical novel of 500,000 characters could be shared in 9 posts.

This evolution of the *post* concept corresponded to a subtle but significant change in Facebook's *profile* concept. When Facebook was first deployed, the intent was that users would visit each other's profile pages, and look at the information and photos placed on them. Users notified each other of changes to their profiles through *status updates*. These status updates, which later came to be called posts, grew in size and significance and became the content themselves. Now when you visit a Facebook user's profile, the main attraction is no longer the *profile* but the *timeline*, which offers a chronological sequence of the user's posts.

Concepts are purposive. Concepts have purposes; that's why they were invented. The purpose of *trash* is to allow undoing of deletions; the purpose of *style* is to help maintain consistent formatting; the purpose of *reservation* is to achieve efficient use of resources.

If you can't identify a compelling purpose, you probably don't have a concept. This will mean that sometimes apparently central notions in a software design turn out not to be concepts. The user of a system is not a concept; if there's data associated with users, it's probably there for concepts that have easily identifiable purposes, such as *authentication* (identify users of a resource) and *account* (aggregate the transactions of a single user).

Novices often assume that entities that are important in the real world context of a design must be concepts. If you're designing a software system for a bank, you might imagine that *bank* itself should be a concept. But you'd almost certainly be wrong. In your data model (and in the code of the system), there may well be *bank* objects. But the structures associated with these objects will likely be in support of other concepts. Most banks have an FDIC certificate number; this is a key identifier for the concept of *fdic insurance*. Likewise, a bank's ABA number is not an inherent property of the bank akin to the number of legs of an animal or the vintage of a wine; it's an identifier that's assigned for a particular role, namely to support the concept of *interbank transfer*.

Of course banks were invented for a purpose, but this purpose is not the reason for including the notion of a bank in the system. This is a rather subtle point. Just because something has a purpose in the world doesn't mean that it becomes a concept in a software system that refers to it. Purposes become relevant only when they are also purposes in the software design itself. So *reservation* is a concept in a restaurant reservation system not just because the concept has a purpose in the external world, but because the concept has the *same* purpose within the system.

To see this distinction more clearly, imagine a system used by banks for managing loans. Clearly such a system will have concepts such as *loan* and *collateral*. But suppose that to determine whether someone is creditworthy the bank collects information about the real estate, bank accounts, and stocks that they hold. Would these be concepts? Almost certainly not. In a stock trading system, *stock* will almost certainly be a concept. But in the loan management system, the purpose of the *stock* concept is not relevant; a borrower's stocks or other holdings are just *assets* that are used to evaluate them. The behavior of stocks in support of their purpose is of no interest in this system; all that matters is that the borrower holds some property whose value fluctuates with time. And so while *stock* would likely not be a relevant concept, *asset* would be; its purpose is for evaluating loan risk (and maybe also to act as collateral).

The same object or collection of objects can thus be viewed as belonging to different concepts depending on the context. A stock holding may be an *equity* in a trading system and an *asset* in a lending system; a book may be a *publication* in a desktop publishing system and a *holding* in a library system.

This is not a new idea. In the 1970s, the Turing Award winner Charles Bachman developed the *role model*, an extension to the network database model that augmented entities with "role segments." A role, according to Bachman, is a "behavior pattern which may be assumed by entities of different kinds" [7]. Classifying things according to their *behavior* is likewise the essence of concepts. The idea of roles has appeared repeatedly in programming and software development since then. It motivates the notion of cross-cutting types defined by their operations (such as Java's interfaces), and is the focus of an entire development method known as Object Oriented Role Analysis and Modeling (OOram) [128].

Concepts are behavioral. The concepts we've looked at have both static structure (in their state) and dynamic behavior (in their actions), but it's behavior that really defines a concept. To the user of the *trash* concept, the fact that the trash contains items is secondary to the more important fact that those items can be restored or permanently removed. To the user of the *style* concept, what matters is that if you change a style, all paragraphs that have been assigned that style are updated in concert; that the word processor must remember the style of each paragraph is just a prerequisite to this important behavior. And finally, the restaurant customer wants to be able to book a table, turn up and find it available; that the restaurant chooses to use a reservation book is of no interest to the customer.

This quality of concepts leads to a simple rule of thumb: if there's no behavior, there's no concept. If you're designing a photo editing app, you might be tempted to identify *pixel* as a concept. It's certainly a concept in the informal sense of the word. But what behavior is associated with pixels? Asking this question moves you towards the real design concepts. Perhaps the behavior is that you can take a photo comprising an array of pixels and edit it by making all the pixels darker or lighter; in that case, the concept might be *adjustment*. Or perhaps the behavior is that you can split a pixel into its constituent red, blue and green components, and edit just one of those components for all the pixels in the image; in that case, the concept might be *channel*. Or the behavior is building up an image by combining pixels from multiple overlapping pixel arrays; then the concept might be *layer*.

Identifying behavior first is useful from a design point of view because without behavior there's very little to design. All the complexity of a software app comes from its behavior. Suppose you're designing a photo library in which each photo has a collection of metadata fields that include camera settings, capture time, location, and so on. Storing and retrieving this metadata will obviously be important, but our first question should not be: what metadata fields are there and what is their structure? Rather, we should ask: what do people do with metadata? Can the metadata of a photo be changed? Can you change the capture time of a photo? Can the metadata be augmented with user-specific fields? What kind of sorting and searching can you do? When you share photos, can you erase metadata for privacy? These are the essential questions that will impact the design of the photo library, and will likely lead to a richer concept of *metadata* than if we just assumed that the metadata of a photo was a static structure amenable to occasional updates.

Concepts have rich states. The structure of a concept's state, although secondary to its dynamic behavior, nevertheless plays a vital role. A trivial state structure may be evidence that a concept is degenerate and not really a concept at all.

Looking for state structure, as with behavior, can often point us towards the real concepts. If the concept of *password* comprised just the password itself, along with some criteria for not being easily guessed, we wouldn't have much of a concept. But once we include the association between accounts and their passwords and usernames, we are moving towards a useful concept of *authentication*.

Occasionally, the state structure of a concept reveals something essential. In the *style* concept (see Figure 4.4 and Note 45), maintaining an association between elements and their styles is the key to propagating style changes back to elements. Without it, you have the "pseudo styles" of TextEdit that don't allow you to make changes to elements after they've been formatted without reapplying a style to each element.

This is a nice application of an adage due to David Wheeler, a computer science pioneer at Cambridge University who worked on the original EDSAC computer and wrote the first computer program ever to be stored in a computer's working memory. Wheeler said "There is no problem in computer science that cannot be solved by introducing another level of indirection." The *style* concept achieves its effect—allowing multiple elements to have their format updated at once—by associating formats not directly with elements, but indirectly, through styles. In an email exchange that I had with Wheeler in 2001, he confirmed that he was the author of this adage, but noted that he usually followed it by something like "and this usually reveals new problems." (Another wise aphorism of his: "Compatibility means deliberately repeating other people's mistakes.")

I noted another example earlier of state structure revealing essential design questions: the *upvote* concept appears to most naive users to just associate a count—the number of votes—with each item. But to prevent double voting in which a user upvotes the same comment in a forum multiple times, say, the concept must track not only the number of votes but also *who* voted for each item. Representing "who" becomes an interesting and difficult design question. To track users by their usernames, you'd need to have users be authenticated before they can vote; tracking them by an IP address gets around this, but would allow a user to vote twice by switching to a different network; using the browser identifier overcomes that problem, but now allows a user to vote twice if they have a second browser; and so on.

Concepts localize data models. A big idea in software development, from the earliest days of databases, is to start with a *data model* or *domain model* that reflects the subject matter of the system being built. So if you were building software for a library, you might model the card catalog, the shelf layouts, the borrowers, and so on.

The first benefit of a domain model is that it provides a single structure and vocabulary that can be shared by all programmers, so you can avoid the common situation in which two programmers develop different and conflicting models of the same thing. A second benefit is that the domain model, by reflecting essential and unchanging aspects of the problem, provides a stable foundation for evolution: even as functions are added, the underlying domain doesn't change. Grounding a directions-finding app in a domain model of roads and intersections, for example, might allow you to evolve the app without having to make frequent changes to the model, since the layout patterns for roads have been fixed for decades (in the US at least since the 1950s), and only change occasionally and in small respects (for example, with the advent of high occupancy lanes in the 1980s and 1990s).

On the other hand, the domain model may tempt you to include too much data. Should your library model include the kind of paper each book is printed on? The color of the cover? It depends, of course, on what functionality you intend to support. If the library is for book collectors, the kind of paper will be important. And it turns out that most readers recall books by the color of their covers, and the lack of this information in traditional library catalogs leads some librarians to use Google search for finding books (since it allows you to search for images by color).

Basing a design on concepts encourages you to do just the right amount of domain modeling. Each concept contains in its state definition whatever domain model structures are needed for that concept's behavior. It's always clear if the structure has more or less detail than the behavior requires. The domain model as a whole becomes just the composition of the state structures of all the concepts. By separating the model into smaller and clearly motivated structures, the role that each structure plays is evident, and you can grow or shrink the data model depending on which concepts are needed.

More subtly, as we'll see in Chapter 6, concepts also let you factor out different aspects of a complex behavior. For example, in a conventional data model of a help forum, you might associate a post both with its content and with the users who have registered interest in it. These two associations belong to different concepts, however: the content association belongs to the *post* concept itself, but the user association is part of a *notification* concept.

Concepts are generic. You might have noticed that, in describing the three concepts of Chapter 4, there was a discrepancy between how the concepts were introduced and motivated and how they were defined in the figures. The *trash* concept was explained as a way to manage files; the *style* concept was for associating formats with paragraphs; and the *reservation* concept was for booking tables at a restaurant or books in a library. But in the figures, *trash* was about deleting and restoring "items"; *style* for associating "formats" with "elements"; and *reservation* for booking "resources." These types—listed after the name of the concept—are not real types but placeholders (or what computer scientists would call "type variables") for the actual types that will replace them in a real usage.

When we encounter a concept for the first time, it's always in some specific context, and some concepts (such as *style*) are easier to understand in concrete terms. (Starting with the concrete and moving to the abstract is a standard pedagogical strategy known as "concreteness fading" [21].)

But almost all concepts are generic, and can be applied to different types of objects according to the context. So the *trash* concept handles deletion of files in the Macintosh Finder but messages in Gmail. The *style* concept is generic in both the format and element types, so it can be used to apply colors to graphical objects in InDesign, or text formats to paragraphs in Word. As revealed by its type parameters, the *reservation* concept is generic not only in the resources but also in the users. In a railway network, to prevent trains from occupying the same track segments at the same time, a reservation concept can be applied: in this case, the resources are the segments and the users are the trains.

(In the design of the Macintosh *trash*, folder structure does actually play a role. But this structure does not affect the essential behavior of the concept; more significant than the fact that you can delete folders is that the trash *itself* is represented as a folder. For this reason, I see this not as a violation of the generic nature of the *trash* concept, but instead as a very skillful synergy of two distinct concepts, *trash* and *folder*. See Chapter 6.)

Not all concepts can be generic. Although the *style* concept is generic, treating elements and their formats as having no specific properties, the elements and formats themselves will be very specific in any particular app, and therefore provided by more concrete concepts. So in a word processor, *paragraph* and *format* will be highly specific to word processing. Separating the generic and non-generic concepts also helps make the code cleaner and more flexible.

By viewing a concept in its generic form, you can strip away accidental details and grasp its essence more clearly. Sometimes this is easy to do. A user already accustomed to upvoting content on a social media apps will grasp the idea immediately when encountering it elsewhere (for example, in the comments of a newspaper). Sometimes the generalization is harder, but the rewards are greater: once you realize that Apple Keynote, for example, offers styles not only for paragraphs but also for shapes, you see new possibilities for saving work and making your presentations more consistent.

Here's a powerful design exercise. When you're working on a concept, look for respects in which it might fail to be generic, and ask yourself if you can adjust the design to achieve full genericity. Suppose you were looking at the *style* concept, for example, in the context of a word processor. Your first reaction might be to think that there's nothing generic about the concept at all, since the dialogs for defining styles contain so many typographic details. Then you might realize that these details all belong to a different concept, namely *format*, and that the style dialogs are just dominated by format features. With this insight, you're now motivated to check that *style* and *format* are cleanly separated conceptually, even if their user interfaces overlap. Are all the format settings that are available within the style dialogs separately available for direct formatting, and vice versa? And if you can cut-and-paste formats, are all the same format settings available for that too?

In Adobe InDesign, for example, the format settings that can be applied directly to paragraphs are the very same settings available for paragraph styles—even though, disconcertingly, the two have very different user interface dialogs (and even when they share a menu, the menu items appear in different orders). I have found one non-generic aspect of InDesign's paragraph style concept: its *next style* feature, which lets you set a default for the style of the next paragraph. This relies on the fact that paragraphs appear in sequence, and introduces a tricky coupling between concepts. To find the style of a paragraph, the *style* concept alone is not enough; if the paragraph has no assigned style, you need to check to see if its predecessor has an assigned style.

This is not necessarily bad. The next style feature is certainly useful. But the principle that concepts should be generic whenever possible highlights the price paid for this benefit: a coupling that I suspect introduces some complexity in the code.

Concepts are freestanding. Perhaps the most significant quality of concepts, and the one that makes them a good basis for software design, is their mutual independence. Each concept is freestanding, defined by itself without reference to other concepts.

So our *style* concept, while mentioning elements and their formats, doesn't depend on their properties in any way. The *trash* concept doesn't depend on any properties of the items being deleted. And the *reservation* concept doesn't depend on the resources being reserved.

A key tenet of programming is to make the code modules as independent of each other as possible. All the reasons for doing this in code apply equally to concepts. If the concepts are independent, they can be worked on independently; a designer working on one concept can't have their work undone by a mistake or assumption made by a designer working on another. The design work itself becomes easier, because you only need to think about the concept you're working on, and not how it might interact with others. And if concepts are independent, you have the flexibility to add and remove concepts, and to create variants of the design containing subsets of the concepts.

Concept independence is essential for incremental design. When you're faced with the design of an entire app, you need some way to get started, to break the design problem into pieces. If you break the design into the wrong pieces, you'll find that you can't work on them one at a time, because when you adjust one it breaks another, like a game of whack-a-mole. By identifying the concepts, you can approach a design piecemeal, even if you find yourself revising the overall set of concepts (which is, of course, inevitable).

Sometimes, in working on the design of a concept, you may find yourself drawn down complicated side alleys. For example, in the design of a restaurant reservation concept, you might want to say that a notification is sent to the customer after a reservation has been accepted. This risks taking you down a rabbit hole into questions of how you notify the customer, what happens if the notification fails, and so on. A much better approach is to treat *notification* as its own concept, and to design it and the *reservation* concept completely independently of one another. Then, with the two concepts in hand, you can ask yourself: which actions in a reservation should produce notifications? Not only does this neatly separate concerns, but it also leads to a more flexible implementation. (How concepts are joined together in this way is explained in Chapter 6.)

When concepts are composed in an application, dependencies arise in the sense that including one concept in the design may make sense only if another concept is included too. For example, in a social media app, the *comment* concept may depend on the *post* concept, because comments only make sense if there is something to comment on. But this form of dependence, which is explained in Chapter 7, does not vitiate the fundamental freestanding nature of concepts and their independence, since the dependencies belong to the context of use (that is, the app) and not to the concepts themselves.

How concepts differ from abstract types and objects. Abstract types (or abstract datatypes, ADTs) are software modules that provide a collection of objects along with associated functions. Two related but distinct concerns motivated their invention. First was the realization that programming languages came with a collection of built-in datatypes, such as integers and strings, but there was no way to add new datatypes. If your program manipulated vectors, for example, it seemed unreasonable that adding and subtracting vectors should be so different from adding and subtracting integers. A second concern was that complex data structures often became obstacles to change: so many parts of the program depended on the exact form of the data structure that any adjustment to it would ripple across the whole program. The solution in both cases was the idea of the abstract type: a module that provided a collection of values, and operations that could be performed on those values, but which hid the way in which the values were represented.

Abstract types differ from concepts in two key respects. First, abstract types are most potent when they embody collections of immutable values, such as the pixels of a photo editing app or the email addresses of an email client. These are generally *not* concepts, because the behaviors that they support are associated with larger structures: pixels are good for image editing, but that requires pixelated images or channels, and email addresses are good for communication, which requires messages.

Second, abstract types—like the objects of object-oriented programming—are a code-level and not a design-level notion. Consider, for example, the concept of *comment*. This concept would encapsulate the behavior of creating, editing, and deleting comments, and its state would associate comments with their targets (which might be social media posts, or newspaper articles, etc.). In code built with abstract types, we could introduce a *comment* type with operations for creating and editing comments, but connecting comments to their targets would be hard. The *comment* values themselves might point to their targets, but this would not allow you to find the relevant comments given a target. The target type itself, say *post*, could contain a reference to the *comment* datatype, but that would be undesirable since it would introduce a dependence of *post* on *comment*, violating the Parnas criterion discussed in Note 81 (namely that if *post* depends on *comment* there should be no useful subset that includes *post* but not *comment*).

The conventional solution to this dilemma is to build a mapping from targets to comments. From a concept design perspective, this mapping belongs to the *comment* concept. In a program structured with abstract types, it would probably be implemented as its own type. Some programmers might place such a mapping inside the *comment* type, but the resulting module would hardly be an abstract type any longer. And even as an object it would violate an oft-cited principle of object-oriented programming that objects should not have static components.

How concepts differ from features. A "feature" is an increment of functionality. A family of products may be described by a set of possible features, with each member of the family characterized by those features it possesses. Sometimes the term is used for a large aggregation of functions (as in the "photo cataloging feature" of Adobe Lightroom), but it may also represent something smaller (as in the "automatic page numbering feature" of Microsoft Word).

One of the best known frameworks for software development using features is Don Batory's GenVoca [8]. Each feature is viewed as a transformation on the program that augments it by adding new elements; the construction of a system can then be understood as a sequence of transformations starting with an empty program (which might be regarded as the most basic feature!) and ending with the final version of the program. A product family can be viewed as a set of features, with all the possible programs that can be obtained from them. This framework makes it easy to manage such a family, and can even allow the code for different feature combinations to be generated utomatically.

A concept might be viewed as a kind of feature, in the sense that it brings an increment of functionality to an app. But the converse is not the case: not every feature is a concept. Features are rarely freestanding, and, unlike concepts, are not designed to be used in different product families and domains. Instead, the features of a family are more akin to a domain-specific language designed expressly for that family. Features therefore do not generally have their own purposes, but rather are characterized by the way in which they extend the functionality of other features, layered one on top of another.

Features, in Batory's sense, unlike concepts, are not primarily user-facing, so a user need not be aware of the feature boundaries. Whether some functionality is added in one or two features, for example, affects only the space of programs that can be built; for concepts, in contrast, the identity of a single concept in the user's mind is essential. Moreover, one feature can replace or modify the behavior of another feature, whereas concepts always retain their behavior when composed (Chapter 11).

The term "feature" is used somewhat differently in the context of telecommunications to refer to the variety of services—automatic callback, call forwarding, call waiting, and so on—offered by a system that can be sold as distinct products to customers, and configured for them individually. These features, unlike Batory's features, are very much user-facing, and are implemented not as code transformations but as separate modules.

A major concern of such features is "feature interaction": the problem that features may behave unexpectedly when combined. The fix is usually either to disallow certain combinations, or to introduce modifications of behavior (such as precedence rules that favor one feature over another). Glenn Bruns provides a helpful classification of feature interactions that uses a simple formalism to clarify the main distinctions [22]. I'll return to feature interaction when we look at integrity, its analogue for concepts (Note 110).

(Thanks to Don Batory for help preparing this note.)

How concepts differ from mental concepts. Mental concepts exist in the mind alone. When user interface designers talk about the mental models of users, they raise the issue precisely because these mental models often differ from the designer's intent. A naive user of a thermostat may have a mental model that suggests that turning the thermostat up or down more will cause faster heating or cooling. And many people, it seems, think that pressing elevator buttons multiple times makes them come faster. People use the term "concept" to describe the mental images that these models comprise as distinct from the reality. In this book's usage of the term, however, concepts are very real—and when they are well designed, generate mental images that correspond exactly to the reality.

The relationship between concepts and the field of conceptual modeling is discussed at length in Note 14.

Chapter 5: Concept Purposes

49. **Purposes in design thinking: needfinding**. In the design of software, the question of purpose arises before concepts are considered with the larger question of what the entire system or app is for. That question has been extensively explored by researchers in different fields under different names.

In the field of design, the term used is "needfinding," and generally involves empirical investigations of potential users, which might include interviews, observation, and ethnographic studies. The term was probably coined by Robert McKim in the 1960s, who argued that designers needed to be involved earlier in the process if they were to have more influence. (McKim, incidentally, became a mentor to David Kelley, a founder of the design company IDEO and Stanford University's d.school, and a leading proponent of design thinking.)

In 1984, Rolf Faste joined the faculty of mechanical engineering at Stanford, took over from McKim as the director of the Design Program, and continued his work in needfinding. In an entertaining and insightful paper [39], he outlined the needfinding techniques that he taught students in the program, as well as a list of "blocks to perceiving needs" that included a recognition that needfinding, while being a critical and creative act, is usually undervalued. In a wry philosophical observation, Faste noted: "The need itself is a perceived lack, something that is missing. Needfinding is thus a paradoxical activity—what is sought is a circumstance where something is missing. In order to find and articulate a need, this missing thing must be seen and recognized by someone."

Purposes in software engineering: requirements. In the field of software development, the term used is "requirements engineering," also the title of a major annual conference. Requirements engineering has placed less emphasis on the experiences of individual users, recognizing that in many systems, the key requirements may be associated with "stakeholders" who are not always users themselves. Indeed, in many such systems, users and operators are best treated as components of the designed system that fulfill requirements in collaboration with software and hardware. The requirements of a traffic light system, for example, are to achieve safe and efficient flow of traffic, and the individual drivers are relevant only to the extent that the design must account for their behavior in ensuring that the requirements are met.

A coherent theory of software requirements was developed by Michael Jackson and Pamela Zave in a collaboration in the 1990s that began by focusing on telephone switches (5ESS in particular) but soon expanded to software systems in general [155]. Their ideas were expanded in Jackson's book, *Software Requirements and Specifications* [69], and crystallized in a reference model developed with Carl Gunter and Elsa Gunter [49]. The central tenet of this work is that requirements are not merely "high-level specifications," and often cannot be expressed in terms of inputs and outputs of the system being built. In research circles (especially in requirements engineering), these ideas are now widely understood, but in industry a more primitive view of requirements may still obtain. It is not uncommon to hear of systems that have "thousands of requirements," which turn out not to be requirements at all, but rather fragments of interface behavior. Thus "the system shall terminate immediately when the emergency stop button is pressed" might purport to be a "requirement," but actually is no more than a partial specification of a system function. Formulated in this way, it fails to address the real requirements issues, such as how pressing the stop button might impact ongoing processes in physical peripherals.

Another leading thinker in requirements engineering is Axel van Lamsweerde [89], who developed an approach based on goals [87] that can be formulated as temporal logic properties, can be checked systematically against the behavior of agents (which include system components, users, operators, etc.), and can even be used to synthesize software. An interesting aspect of the goal-driven approach is the identification of "obstacles" [127, 88], which are unanticipated behaviors that prevent goals from being fulfilled (and which are typically resolved either by changing the agents or by narrowing the goal).

Concept design can benefit from almost all the ideas that have been developed around requirements and needfinding, with two shifts of perspective. First, requirements are not formulated monolithically, but are broken down into the purposes of individual concepts (much as goals are assigned to agents in Lamsweerde's approach). Second, before exploring a purpose, the concept designer looks to see if a concept already exists that might serve it, hoping that the existing concept will embody the accumulated knowledge not only of its design but also its requirements.

50. **Beneficent difficulty**. Michael Jackson coined the term *beneficent difficulty* to capture his insight that design methods that make things too easy are suspect [69]. Encountering difficulties can be a positive sign that you are really engaging with the problem and making progress.

51. **Concept metaphors aren't helpful**. One common way to explain concepts is to appeal to metaphors. But these are more often than not misleading. In explaining the *layer* concept in Adobe Photoshop, for example, people often liken layers to the acetate sheets that graphic designers used in the past to build up an image layer by layer. But this metaphor misses the point. Acetate sheets were used for color separations (which in Photoshop are done with channels, not layers), and for animation. The purpose of the Photoshop *layer* concept is very different: it's to allow *non-destructive editing*. You can create a layer that darkens an image, say, and modify it later, increasing or decreasing the effect, or even throw it away, without ever

modifying the pixels of the image itself (and having to rely on old versions or undo if you made a mistake). Until you understand that, you won't be able to understand layers or use them effectively. (See also Note 40 for why the Apple trash is not like your kitchen trash can.)

This is not to say, by the way, that software concepts should never be explained in terms of real-world concepts—but that's usually because they're the same concept. The *reservation* concept in the OpenTable restaurant reservation app isn't *like* the restaurant reservation concept in which you call on the phone and someone writes your name in a book—it *is* that concept! The only difference is the hardware on which the concepts are executed (a computer for one and a notebook for the other).

52. **Secrets of call forwarding**. The analysis of the call forwarding problem is due to Pamela Zave [154], whose work has helped clarify many fundamental questions about the design of communication systems, and which—unusually—combines powerful use of formal methods with deep attention to conceptual issues.

53. **The mysterious Facebook poke**. The lack of a compelling purpose for the *poke* concept was part of its allure. According to an article in Slate [148] the Facebook help page originally said: "When we created the poke, we thought it would be cool to have a feature without any specific purpose."

According to Urban Dictionary, the poke "allows users to say hello to or show interest in a friend without having to go through the tedious process of crafting coherent sentences."

That same Slate article investigated the purpose of the *poke* by interviewing users to find out what they did with it. The author was evidently not convinced, and concluded: "By now it's clear that the Age of the Poke is behind us." Facebook seemed to lose confidence itself, and buried the poke button deep in menus.

But then, thrillingly, a few years later (in 2017), the poke button made a proud return in a prominent position. Now (in 2021), poking has been exiled again, and as far as I can tell, is not reachable through any menu. But it has been resurrected at a special URL with its own page that offers "suggested pokes" and allows you to search for a friend to poke. Thank goodness!

54. **Norman's refrigerator**. A more compelling, but more complicated physical example of concepts without purposes is given by Don Norman [110]. Many refrigerators have two separate controls, one marked "fresh food" and one marked "freezer." The labels give the impression that these controls set the temperatures in the two

compartments, and can be adjusted independently. In fact, one controls the compressor and the other adjusts the ratio of cold air between compartments!

As with the old faucet, there is no alignment between the user's purposes and the concepts that are provided. If you want to make the fresh food compartment colder, you can turn down that setting, but it will affect the temperature of the freezer compartment too. And, like the faucet, a series of adjustments of both controls is needed to converge on the desired state, but in this case it's even worse because the temperatures change slowly, and the adjustments must be made over a period of several days.

Where does the difficulty reside? Norman's diagnosis of the refrigerator's design problem is subtly different from mine. His primary concern is that the refrigerator does not convey the conceptual model effectively, and so the user's mental model and the actual model are inconsistent. My objection, in contrast, focuses on the conceptual model itself: the design would still be bad even if the controls were accurately labeled.

To Norman, operating the refrigerator is an example of the "surprisingly large number of everyday tasks" for which "the difficulty resides entirely in deriving the relationships between the mental intentions and interpretations and the physical actions and states." His language is revealing, and points to his origins as a psychologist: indeed, Norman's book [110] was originally titled *The Psychology of Everyday Things* and was only later renamed *The Design of Everyday Things*. The incompatibility he sees is between a physical world (that seems immutable the way he describes it) and the more malleable world of "mental intentions and interpretations," which can be shaped by the user interface.

In my analysis, while the user's difficulty may reside in the psychological gulf between mental model and physical mechanism, its resolution resides somewhere else: in the design of the refrigerator's concepts, which are simply not fit for purpose. The user's purposes—adjusting the temperatures of the fresh and frozen compartments—exist independently of any mental intention, and it is the fundamental misalignment of purposes and concepts that makes the design bad. This is not to say, of course, that Norman fails to recognize the troubling design of the refrigerator itself, and wouldn't prefer a design with better concepts, but that his work has focused more on identifying the psychological barriers to usability.

Gulfs of execution and evaluation. An influential paper [60] by Edwin Hutchins, James Hollan and Donald Norman described the distance between the user's mental model and the actual conceptual model in terms of two "gulfs." The Gulf of Execution separates the intentions of the user from the provided actions of the system; and the Gulf of Evaluation reflects the effort the user must exert to interpret the state of the system and determine how well the intentions have been met.

Both gulfs are evident in the refrigerator example: the intention to lower the temperature in the fresh food compartment but not the freezer has no directly corresponding action, and the effects of adjustments cannot even be observed until many hours later when the temperatures have stabilized.

The gulfs helpfully identify certain common difficulties that users face, and sometimes suggest immediate remedies. Providing an intuitive action that corresponds exactly to the user's intent may eliminate the Gulf of Execution, for example. In pre-graphical operating systems, moving a file between directories meant knowing the syntax of the move command and dealing with absolute and relative paths; in the WIMP (windows-icon-menus-pointer) interface, moving a file between folders became a literal move, with the user dragging the file icon from one window to another.

In other cases, however, an effective design may require *teaching* the user to perform an action that may seem unintuitive at first. In some early car radios, to assign a radio station to a preset button, you had to follow a series of steps. In the better designed devices, instructions appeared on the LCD display, seemingly shrinking the Gulfs of Execution and Evaluation. But the design was actually terrible, because it was so inconvenient and tedious to modify the presets, and, worse, one was tempted to do it while driving. A much better solution emerged. To assign the currently playing station to a preset, you simply hold down the desired preset button until it beeps—hardly an action that comes naturally.

In other cases, a better design might change the set of expected actions. Suppose you want to update the formatting of many of the paragraphs in a document, say changing all the section headers to bold. You could select each paragraph in turn, and apply the new formatting to it. Noting how laborious this is, you might complain that there is no easy way to select all the paragraphs you want first, and then apply the formatting to them together. In other words, you experience a Gulf of Execution between your mental intention (select an arbitrary collection of paragraphs) and what the application offers (the ability only to select one at a time).

A good solution to this problem, as we've seen in Chapter 4, is not to bridge the gulf for this particular intent, but rather to introduce a concept (*style*) that fulfills the user's purpose, albeit by requiring new behaviors and mental intentions (involving defining styles and pre-assigning them to paragraphs).

In summary, the notion of gulfs is invaluable for exposing usability snags, but in discovering a gulf between a user intention and a system action, one should not regard either the intention or the action as given, and the best design solution may involve changing both.

55. **Purposeless concepts in Git.** The *editor buffer* example is nice because it shows a concept lacking a compelling purpose (even if it had a rationale for the implementor). But the price paid for the lack of purpose was merely some needless complexity (and a risk of losing data prior to the buffer being saved to file).

A more compelling example is provided by Git, a widely used version control system that is known both for its power and for the complexity of its design. Many of the foundational concepts of Git appear to have no intelligible purpose. Some appear to exist only as workarounds for design flaws; *stash*, for example, compensates for problems with *branch*.

Others expose the underlying mechanism but don't fulfill any significant purpose, even if they can be put to use in various ways. In particular, the *index* (also called the *staging area*) sits between the working area and the repository and *almost* provides tracking of files, selective commit, and reversion to old file states. But it does none of these things fully or consistently, and is a stumbling block for novices, in large part because none of the introductory materials explain what this concept is *for*.

It was studying Git, in fact, that led to the realization that purposes were crucial to understanding and designing concepts. My student Santiago Perez De Rosso performed an analysis of Git's design using my emergent theory of concepts, and developed a conceptual redesign called Gitless that eliminates many of Git's usability problems by reshaping the concepts to align more closely with user purposes [118]. When we were trying to understand the *index* concept, we read many online explanations and were struck by the fact that nobody really knew exactly what it was for. And to the extent that they did, each explanation proposed a different purpose, often incompatible with the others! This led not only to the idea that purposes matter, but also that claiming multiple purposes for a single concept is problematic (as explained in Chapter 9).

Purposeless concepts in programming languages. Examples of purposeless concepts introduced as workarounds can be found also in programming languages. The *boxing* concept of the Java programming language, for example, makes up for a spurious distinction between primitive values and objects (which had been eliminated in the design of CLU in 1977!).

Likewise the complicated and confusing concept of *this object* in JavaScript, which introduces a strange form of dynamic scoping at variance with the lexical scoping of all other identifiers, was motivated by the desire to support an object-oriented programming style, even though the dynamic scoping that *this object* introduces has nothing to do with object-oriented programming (and creates new problems of its own).

56. **How to find the old "save as" command in the Apple file menu**. Many users were unhappy about this change, not because they missed the buffer, but because they preferred *save as* to duplicating and renaming. Responding to complaints, Apple returned "save as" as an optional command (in os x Mountain Lion), accessed by holding down the option key while opening the menu.

57. **Stars, hearts and Twitter games**. Twitter explained in an online announcement in 2015: "We are changing our star icon for favorites to a heart and we'll be calling them likes. We want to make Twitter easier and more rewarding to use, and we know that at times the star could be confusing, especially to newcomers. You might like a lot of things, but not everything can be your favorite."

This completely failed to address the question of what the concept of *favorite* was actually *for*. Twitter adjusted the design at the linguistic level (Chapter 2) but the problem was in the conceptual level.

In fact, some expert Twitter users had already noted the unclear purpose of the concept, and had found their own purposes. Matthew Ingram explained: "The problem for Twitter is that the 'favorite' function had developed a range of uses over time, many of which are known only to the journalists and social-media experts who spend all their time on the service." Ingram went on to say that he used the concept precisely in the way Melania Trump intended, apparently unconcerned about the results being public: "For some (including me), clicking the star icon was a way of saving a tweet for later, or of sending a link that was being shared to a service like Instapaper or Pocket."

Another blogger, Casey Newton, reported a completely different purpose that exploited its public nature: "I've favorited more than 60,000 tweets over the years, and in that time I've come to appreciate how versatile that little button is. I use it as a kind of read receipt to acknowledge replies; I use it whenever a tweet makes me laugh out loud; I use it when someone criticizes me by name in the hopes that seeing it's one of my 'favorite' tweets will confuse and upset them."

Perhaps most revealing was a comment made by Chris Saca, a Twitter board member, explaining the change from a star to a heart: "If Twitter integrated a simple heart gesture into each Tweet, engagement across the entire service would explode. More of us would be getting loving feedback on our posts and that would directly encourage more posting and more frequent visits to Twitter."

58. **New concepts for image sizing**. Here's how I might apply the idea of *purposes* to the redesign of *image size* and related concepts for photos. I'd start by identifying the most pressing purpose, which seems to be: downsizing photos to meet quality and space constraints.

Then I might start imagining an operational principle. Maybe I'd ask the user what physical size the image is to be displayed at (three inches across for a phone, or three feet across for a gallery wall?), and would show a sample of the image at that size with an indication of whether its quality is sufficient, insufficient, or even excessive for this size. I'd also show the file size, with an indication of its suitability for different uses (sending in email, displaying on a web page, etc.). You'd be able to choose a use, or a particular target file size in megabytes, and the display would be updated to show you the new quality. Once satisfied with the balance of quality and file size, you would export the image to a file.

This exploration is focused on a single purpose. Other purposes might be valuable to some users too, such as upsampling images (interpolating new pixels so they can be printed at larger sizes) and setting default printing or display size. My contention is that bringing in all these purposes was responsible for the complexity of the Photoshop design, and that limiting the design exploration to just the one basic purpose (of downsizing) would likely lead to a more robust and usable concept, allowing the other purposes to be addressed in an expert mode or in concepts of their own.

Sizes in CSS: a complicated tale. The *image size* concept might seem obscure, but once you understand the purpose (setting a default size for physical printing and layout), it makes sense. And the operational principle is trivial: if you set the image size of a photo to be 4 × 6 inches, and you then print it (without scaling it or changing the size), the physical size of the photo will be 4 × 6 inches too. It's a nice case of the operational principle being easier to grasp than the purpose.

The size concept in CSS (Cascading Style Sheets, the layout language for web pages) is baffling in comparison. The following explanation is taken from an anonymous blog [113].

The concept's complications arise from two very real problems. The first is one that you may have thought about if you gave a slide presentation and were asked to use fonts "no smaller than 18pt." What on earth does that mean? It's clear what 18 points means on the printed page—it's about a quarter of an inch. That's presumably not the intended size on the screen, and slides themselves don't have physical dimensions. In practice, what it means of course is "choose a type size of 18pt in your slide presentation app," but that just makes the problem someone else's (and may not work if you use a different app). This suggests that the meaning of "18pt" should be display-dependent, and correspond to a larger size on a larger display.

The second problem arises with pixel measurements. Obviously "18 pixels" can't literally mean a distance that covers 18 physical pixels in the display, because that would mean that if you bought a new phone with double the screen resolution, the text in your browser would halve in size. On the other hand, pixels can't be converted into inches (say), because you wouldn't want a one pixel border to turn out to have a width that is a non-integral number of pixels (and produce a blurry line).

The CSS standard resolves these issues as follows. It defines a "reference pixel," which is a pixel on a 96 dots per inch monitor positioned at arms length from the viewer (the typical web viewing setup from the 1990s). It recommends that the designer of the "user agent" (the app, such as a browser, that is going to translate the measurements into commands for a physical display) select a size for a one-pixel measurement that corresponds to an integral number of device pixels that will produce the same *apparent size* as the reference pixel, at the expected viewing distance.

This one-pixel measurement that is set in the user agent is called the "anchor unit" because all the other units of measurements are defined from it. If you're familiar with typographic measures, you'll know that one pica is 12 points, and that there are six picas or 72 points to the inch. What you probably did not know is that the anchoring specified in CSS sets one pixel to be equal to 0.75 points.

The upshot of this complicated story is that measurements in CSS work reasonably well. But the operational principle (which I just explained informally) is tricky, and has some surprising implications. In particular, 12pt text on your screen will not be the same size as 12pt text in a book (and an inch on the screen won't correspond to an inch on a ruler); 12pt text on your phone and monitor will have different physical sizes; and a one-pixel border will likely span more than one device pixel.

The practical advice that follows from all of this is straightforward: specify borders in pixels, ignore the display size when you choose your type size, and don't worry about whether you give font sizes in points or pixels (since they're in a fixed ratio).

A more important issue in practice is knowing when to use relative sizes (measured in *em* and *rem*), but that's a story (or an operational principle) for another day.

59. **Chip and PIN**. The weaknesses of chip-and-PIN were reported in [104].

60. **Notes on the Synthesis of Form**. The idea of misfits in this section is drawn from the work of Christopher Alexander, who has been described as the "programmer's favorite architect," mainly due to the influence of his idea of "design patterns" (see Note 104).

    In his book *Notes on the Synthesis of Form* [3], he explains why the idea of "complete" requirements is nonsensical, and how misfits can help: "Such a list of requirements is potentially endless ... But if we think of the requirements from a negative point of view, as potential misfits, there is a simple way of picking a finite set. This is because it is through misfit that the problem originally brings itself to our attention. We take just those relations between form and context which obtrude most strongly, which demand attention most clearly, which seem most likely to go wrong. We cannot do better than this."

    In my own research, I have found Alexander's idea of "fit" (and misfits) to be very compelling and helpful. It crystallizes a key aspect of what makes design hard, especially when addressing security issues or for safety-critical systems. For both, obscure and rarely encountered misfits become central: for security because attackers seek them out and make the rare inevitable, and for safety because rare occurrences cannot be ignored if they are catastrophic.

    Alexander's method for discovering structure. Most of Alexander's book is concerned with a systematic method for discovering structure in requirements. First, all the known potential misfits are enumerated, and expressed as positive requirements. A graph is then constructed in which two misfits are linked if they interact (that is, mitigating one has a positive or negative effect on mitigating the other). Misfits are then grouped so that all the misfits in one grouping interact, and there are no (or few) interactions across groupings. These groupings can furthermore be arranged hierarchically by classifying the misfits into broad areas. With this structure in hand, the designer can then consider design decisions for each group independently, confident that all the factors relevant to a particular decision are accounted for.

    This idea is related to Herb Simon's "nearly decomposable systems" [135]. Alexander never pursued it further, although it influenced others, notably Larry Constantine, who made coupling a key metric in his work on structured design (1975), and maybe also Don Steward, who invented the design structure matrix (1981). Much later, the idea of identifying couplings between requirements was a central element of Eric Yu's i* framework [152] in which links between requirements are marked with a plus or minus to show whether meeting one requirement is likely to make it easier or harder to meet another.

It would be interesting to apply Alexander's decomposition method to concept design, and to see if the clusters of requirements that emerge correspond to concepts (or reflect needed synchronizations between them).

61. **Why verification does not prevent misfits**. Verification, in which a mathematical proof is constructed to show that the code meets its specification, is no better at eliminating misfits than testing, and may even be worse, because misfits correspond to flaws in the *specification* itself.

By broadening the scope of the verification so that the property to be checked is not the specification (that is, the behavior of the system at the interface with the world) but rather the desired *outcome* in the environment, it does become possible to detect misfits—at least those that arise from inadequate consideration of the environment (what Donald Rumsfeld would have classified as "known unknowns"). The outcome relies on assumptions about the environment; simply articulating such assumptions can reveal misfits.

The desired outcome of a traffic light setup at an intersection, for example, might be that cars do not collide; the assumptions might be that drivers obey the lights and that cars cross the intersection at a certain minimum speed. If the traffic light is being designed for an Amish village, a consideration of the minimum speed assumption might reveal the misfit in which the lights change too quickly for horse-drawn carriages.

The structure of such arguments has been a focus of Michael Jackson's work [69, 49, 71] and is central to dependability (or assurance) cases [65, 64].

62. **Blaming the user for accidents**. A fuller story of the Afghanistan PLUGR accident is described in a contemporaneous article in *The Washington Post* [97]. At that point, it was the deadliest "friendly fire" incident of the war.

As inevitably happens when poor design is the real culprit for a lethal accident, investigators looked elsewhere. An Air Force official interviewed for the article said he was unsure whether disciplinary action would be taken, and that he considered the incident "an understandable mistake under the stress of operations." Revealingly, he attributed it to a training problem.

Better training was probably the right workaround in the short term. But the general tendency to cast blame on operators and exculpate designers and manufacturers prevents real progress. The problem is especially acute in medical devices, whose poor designs routinely kill large numbers of patients by overdose and other preventable errors [142].

63. **Catastrophic concept interactions**. The PLGR example illustrates a design flaw not in a single concept but in the interaction *between* concepts. In Chapter 6, I'll explain how concepts are composed and will give many examples of problems that can arise.

The particular case of the *battery* concept is interesting because its interactions with other concepts are linked to potential failures. Ideally, interactions would be minimized—for example, by using persistent storage so that concepts relying on state are not disrupted when power is lost. At the very least, especially for critical devices, it would be wise to consider, for every concept, how its behavior would be affected by intervening battery replacement events.

Concept composition provides a nice framework for ensuring robustness in the presence of troublesome (but necessary) concepts such as *battery*, because it allows the analysis to be factored into a collection of pairwise compositions.

Chapter 6: Concept Composition

64. **Semantics of composition**. In the spirit of Note 44 which introduced a semantics for concepts themselves, here are some notes on how to make precise the composition mechanism described in this chapter.

The concepts of a composition run in parallel, so its traces are just all the possible interleavings of the individual concept traces. The state after a trace of the composition is just the product of the states of each of the individual concepts. The effect of the synchronizations is to further constrain the possible traces. A synchronization of the form

> sync action1 (x)
> action2 (e)

adds a constraint that, for every trace of the composition, every occurrence of the trigger *action1* is followed immediately by an occurrence of the response *action2*. The effect of this constraint is to rule out the interleavings in which the rule is not obeyed. (A note for experts: as formulated, this composition rule makes the trace set no longer prefix-closed.)

The argument *x* of the trigger action is like a quantified variable (or function argument): it gets bound to whatever the actual argument of the action instance is, and then constrains the response action to have the argument given by the expression *e*. In most of the synchronizations I've given, *e* is simply *x*, so the synchronization constrains the actions to have the same arguments.

More generally, a synchronization may be predicated on a condition on the argument to the trigger action (and optionally also on the state of one or more concepts). Thus the synchronization

```
sync label.detach (t, 'pending')
    todo.complete (t)
```

only triggers when the label being detached has the value *'pending'*. The arguments of the response action may be defined not only by the arguments of the trigger action but also by the state. For example,

```
sync email.receive (todo-user, m)
    todo.add (m.content)
```

sets the argument of the response action to be the content of the email message *m* that is obtained by looking up *m* in the *content* relation of the *email* concept. (You might have expected the message *m* to be a composite object with a field called *content*, but all state and structure is localized within the relevant concepts—see *Object mutability* in Note 44.)

The semantics of synchronization is easily extended beyond a single response action to a sequence of multiple actions.

Composition preserves concept behaviors. Suppose we observe an app that is a composition of concepts, looking only at the actions of a particular concept. Since every trace of the composition, by definition, is an interleaving of traces of the individual concepts, the trace we observe must be one of the traces of the original concept. That is, composing concepts *never changes* the behavior of any of the constituent concepts.

This is an essential property of concept composition. It's what makes concepts comprehensible: a concept behaves the same way, irrespective of context. When concepts are not properly composed, or are adapted to work together in a way that undermines their individual specifications, we have a violation of *concept integrity*, which is discussed in Chapter 11.

There is one respect in which a legitimate composition may seem to alter the behavior of a concept. You might interpret a concept specification not only to tell you that certain traces are *not* possible, but rather that certain traces *are* possible. In computer science terms, you might assume that a concept specification asserts not only *safety* but also *liveness*. I have assiduously avoided any such interpretation, because it is important for synchronizations to be able to restrict liveness: just consider an *access control* concept, for example, whose very purpose is to inhibit actions of other concepts when permission to perform them has not been granted.

You may wonder whether asserting in an operational principle that an action can occur vitiates the guarantee that composition preserves specifications. For example, in the *trash* concept (Figure 4.2), the principle included

after delete(x), can restore(x)

which seems to insist that you can always restore an item after it has been deleted. As explained above, however (see *A logic of operational principles* in Note 44), this assertion only requires that the precondition of the *restore* action hold, in terms of the state of the *trash* concept. In other words, the operational principle asserts that the *trash* concept makes the action available, but is consistent with another concept inhibiting it. (Nevertheless, it might be surprising if this happens, and an analysis tool might warn the designer of such cases.)

Generated inputs. Readers familiar with programming languages might have expected the task argument of the *add* action of the *todo* concept (Figure 6.2) to have been treated as an *output* of the action. Adding a task to the todo list involves creating a new task, so it might seem wrong for it to be supplied as an input.

In its most general form, a concept action takes some inputs and the current state of the concept, and produces a new state and an output (or perhaps multiple outputs). The output and the state afterwards are determined by the inputs and the state before. Some inputs come from other concepts. For example, when the action *label.affix (i, l)* occurs, the item *i* is assumed to have been created previously by some other concept. But other inputs can be generated by the concept itself. In the *label.affix (i, l)* action, the label *l* (unlike the item *i*) did not necessarily exist previously. You can think of the label as being selected or formed by the *label* concept, under the direction of the user.

It may help to think about this mechanistically. In an implementation, a concept would include the user interface widgets and code for specifying each input. That input might be a boolean value, defined by a simple checkbox; a short string (for example, one that forms our label input), entered in a text field; a long string, created with a text-editing plug-in; a date selected with a calendar widget; and so on. Each of these inputs, irrespective of complexity, is created with the concept's assistance.

In the *add(t)* action of the *todo* concept, then, the task argument *t* is such an input. In our rather abstract description, we haven't specified what a task actually looks like. In the simplest case, it's just a string. Either way, it plays the role of an *asset* (see *Object roles* in Note 44). A different *todo* concept might separate the name of a task (*t1*, say) from its content ("*file taxes*"). In this case, the task exposed in the *add* action would be a name rather than an asset, but would still be an input created with the concept.

In summary, there are two kinds of inputs to a concept. Some are *generated* with the help of the concept itself; others are provided by other concepts (in a composition). To distinguish these, we can mark the generated inputs with a special keyword. So the *add* action (Figure 6.2) would be declared more fully as

add (gen t: Task)

and likewise the *affix* action of the *label* concept (Figure 6.3) becomes

affix (i: Item, gen l: Label)

In the *email* concept (Figure 6.5), a message and its content are generated by the action that sends it

send (by, for: User, gen m: Message, gen c: Content)

Finally, in the examples of Chapter 4, we would mark the style argument of the action that defines styles in the *style* concept (Figure 4.4)

define (gen s: Style, f: Format)

and the resource argument of the action that registers a new resource in the *reservation* concept (Figure 4.7):

provide (gen r: Resource)

Trace constraints on inputs. In defining composition, I said that the traces of the composite were any interleavings of the traces of the individual concepts, so long as they satisfied the constraints imposed by the synchronizations.

The astute reader might wonder what would forbid a trace starting with the action *label.affix (t, l)* followed by *todo.add (t)*, in which a label is affixed to the task *t before* the task has been created.

The solution to this dilemma is provided by the notion of generated inputs explained in the preceding note. We add a constraint on every trace that every input to an action must be either generated by that action, or generated or output by an earlier action in the trace. In this case, since *t* is not generated in *label.affix (t, l)*, and there is no earlier action to produce it, it cannot be used there. If the actions were reversed, the trace *would* be allowed, because *t* would have been previously generated in *todo.add(t)*.

Origins of composition semantics in CSP. The idea of a composition that preserves the behavior of the individual components is taken from Tony Hoare's CSP (Communicating Sequential Processes) [58]. The theory of concepts as described here differs from CSP in a few respects. First, because concepts are deterministic, CSP's "refusals" (which model the idea that a process might arbitrarily refuse to participate in an event) are not needed. Second, states are observable in concepts, and can be used to constrain synchronizations, but play no role in CSP. Third, concept composition is defined as interleaving, with synchronizations constraining the interleavings with trigger/response pairs; in CSP, processes are synchronized on shared actions.

Concept composition might be cast in CSP terms by representing the synchronizations as a process in their own right, which is then combined with the processes representing the concepts proper. For example, a synchronization in which action *a1* triggers the reaction *a2* might be represented as the process

$SYNC = a1 \rightarrow a2 \rightarrow SYNC \,|\, a2 \rightarrow SYNC$

which allows, repeatedly, either *a1* followed by *a2*, or *a2* by itself.

65. **Bruno Latour's theory of inscription.** The philosopher and sociologist Bruno Latour uses the term "displacement" (as well as "translation" and "shifting") for the way in which a machine may enact a task that was previously performed manually. But far from seeing this as just one aspect of the design of machines, Latour sees it as their very essence, with the users and builders of a machine "inscribed" in its mechanism [90].

Latour's understanding of technology is especially apt for concepts, many of which (the restaurant *reservation*, for example) originate in the world of human protocols and policy and only later become embedded in software. Human concepts that are implemented in software are not a metaphor; they are literal "inscriptions" of a social protocol in code. The misleading claim that the Macintosh trash is a metaphor arises, ironically, not from going too far in relating the virtual to the physical, but in not going far enough and recognizing that it is the very same concept at play (see Note 40).

Dijkstra and anthropomorphism. To Dijkstra, talking about computers as if they were human actors was a cardinal sin [35]. He objected to it because it seemed to undermine the abstract, axiomatic view of software that he preferred, in which code was characterized by its invariants and not by its behaviors. In his defense, invariant-based reasoning was one of the great advances of programming theory, giving programmers a powerful abstraction much like the notion of an orbit in planetary mechanics. But his rejection of operational thinking threw the baby out with the bath water, not only denying us a potent tool for thinking about software, but perhaps missing the very point of what software is about. In concept terms, the state of a concept and its invariants are crucial, but the operational principle is closer to the essence.

66. **Permission based on prior actions**. Another common use of synchronization is to inhibit certain actions in one concept until some actions in another concept have been performed. In the OpenTable restaurant reservation system, you can't post a review (an action of the *review* concept) unless you were previously seated (an action of the *reservation* concept).

    Such synchronizations often suggest the addition of actions to a concept, precisely to allow coordination with another concept. Your *reservation* concept, for example, might not have an action corresponding to a reservation actually being used—in the restaurant case, someone who made a reservation being seated at a table. Even if the action is not strictly necessary to fulfill the purpose of the *reservation* concept itself, it is valuable to other concepts it is likely to be composed with, and therefore worth including (see Figure 4.7). The same rationale applies to including a *no-show* action (as I mentioned in Chapter 4), which allows the reservation system to punish users who routinely make reservations but fail to turn up.

67. **Bridging separated concerns: pages and paragraphs**. Here is a richer, more interesting—and more dangerous—example of collaborative composition being used to bridge between concerns that were separated into distinct concepts.

    In Adobe InDesign, there is a clean separation between the formatting of pages and the formatting of paragraphs. A master page lets you set the dimensions of the text area, and include common elements such as running headers and page numbers. By creating multiple masters, you can customize the layout of each page. In this book, for example, there is a master for each chapter with its own running header, and another one for chapter openings that have no header or page number. At the same time, paragraph styles let you define common typographic formats and apply them consistently.

    The two concepts, *master page* and *style* (explained in detail in Chapter 4) are mostly independent, and this brings clarity and simplicity to the design. Their independence does, however, produce some inconvenience that leads to extra manual work for the user. Whenever I add some text to a chapter in this book, there is a risk that the chapter will spill over onto a new page. If it does so, the last page of the chapter will now be laid out with the chapter-opening master, and the first page of the next chapter will have the chapter-middle master.

    This is not hard to fix. You can add a new page in the middle of the chapter; you can also reassign masters to pages. But both of these require manual intervention. My own preference (which many users of InDesign share) is to edit the text of my document externally; InDesign supports linkage to Microsoft Word for this purpose, as well as to third-party editing tools (although I use a pre-processor that I wrote for myself that allows me to prepare my text in a text editor). This means that you can edit your text externally, and InDesign will detect that the linked text has changed and reimport it. In this context, having to manually adjust pages is a major annoyance.

Mastermatic is an InDesign add-on that addresses precisely this problem. It lets you specify which masters should be used for which paragraph styles. So, for example, you might say that the chapter-opening master should be applied when the page contains a paragraph styled with the chapter-heading style.

This provides welcome automation, but it does not come without costs. The coupling between the *master* and *style* concepts is now rather complicated. What happens if a page includes paragraphs with two different styles that are assigned to different master pages? In this case, Mastermatic resolves the clash by requiring that the style/master bindings be given in a single list, with an implied priority order. I wonder also what would happen if switching to a master with a smaller text box would cause the paragraph whose style dictated the use of that master to spill off the page. I suspect it's for reasons such as this that Adobe did not include this synchronization in InDesign itself.

68. **Subtleties in Gmail's synergistic composition of trash and label**. The designers of Gmail understandably wanted to inhibit the display of deleted messages, so they introduced some ad hoc rules that affect which messages are shown in response to label queries. If your query does not mention the *deleted* label explicitly, it seems to exclude deleted messages from the result; if you mention that label, deleted messages are then no longer suppressed.

This makes sense, although it does violate the integrity principle (Chapter 11) because, strictly speaking, the *label* concept no longer obeys its specification. Gmail does show a warning saying "Some messages in Trash or Spam match your search" with a link to display them, mitigating this problem, but unfortunately, this warning is not shown consistently when messages have been excluded—perhaps an example of trepanning (Note 89).

69. **A synergy trade-off in MIT's Moira app**. Representing administrative groups as if they were mailing lists in Moira is a smart design move, but it's not perfect. Two types of users can be added to mailing lists: MIT users, who have MIT accounts (and thus access to the Moira app itself) and are specified by user names, and external users, who do not have MIT accounts, and are identified by full email addresses. You can include external users in an administrative group too, but it has no effect,

because they can't even log in to the system! This illustrates the kind of trade-offs that synergy may require.

70. **Teabox, synergy and an amusing misfit**. Teabox is a wonderful company that sells Indian teas directly to customers overseas. Their web app employed the synergistic composition I've described, in which free samples are added to the shopping cart as items with zero cost.

This led to a funny anomaly. One time I was purchasing some of my favorite teas, and I entered a coupon code that Teabox had sent me previously. It did not give the discount that I expected. I eventually figured out what was going on. The *coupon* concept used a clever rule that didn't just apply a percentage discount, but rather applied it to a maximum of three items in the shopping cart—and not just any three, but the three with the lowest prices. Two of these items, it turned out, were free samples. The remedy was to remove the free samples from the cart, after which the total cart price dropped precipitously, since two additional and much more expensive items were now being discounted.

This is a nice example of a misfit (Chapter 5) in which a concept itself—in this case, the *coupon* concept—is perfectly plausible, but placed in an unanticipated context has undesirable behaviors.

71. **The remarkable synergies of Photoshop**. Photoshop presents a fascinating example of synergy in concept design. The way this works is undoubtedly complicated, but very instructive from a design perspective. This note might also help those seeking to understand how to use Photoshop more effectively.

In Adobe Photoshop, the central concept, which we might call *pixel array*, involves an array of pixels that can be edited by applying adjustments. An image can be viewed not only in this form as a two-dimensional array of colored pixels, but also as three separate arrays, called *channels*, one for each base color (for example, red, green, and blue).

The *channel* concept allows you to edit an image in each of these colors separately. For example, you might use the blue channel to select pixels that correspond to the sky, or you might boost the darker pixels in the red channel to counter the tendency of shadows to be too cool. Another common use is to split an image into channels in the "Lab" color space, whose dimensions are lightness (L), red-green (a), and blue-yellow (b), and to apply sharpening to the lightness channel alone (which many people believe produces superior results to applying sharpening to the original color image). When editing a channel, the image appears without any

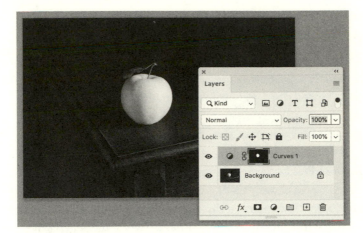

FIG. E.5 *The layer and mask concepts in Photoshop. I've created a layer to whiten the apple in the image; its associated mask ensures that the adjustment is applied only to the apple and not its surroundings (note the white spot in the mask corresponding to the apple).*

colors, since the entire channel only corresponds to a single color. In short, a channel is a gray-scale image.

Another concept in Photoshop is the *selection*, which allows you to select a collection of pixels, and then apply adjustments only to those pixels. What is unusual about this concept, and what distinguishes it from other variants of the *selection* concept (as used, for example, to select a subset of files or email messages for subsequent moving or deletion), is that the selection is not binary: a pixel can be partially selected. When an adjustment is applied, it is attenuated in proportion to the selection: applied in full to the pixels that are fully selected, not applied at all to pixels that are not selected, and applied in part to pixels in between. To make partial selections, the user can apply a brush, with each sweep of the brush increasing the degree of selection (using a feature called "flow"); they also arise naturally when the user defines a selection by drawing a boundary, along with a "feathering" option that produces soft edges so that subsequent adjustments blend in. The essential point is that a selection assigns a value between 0% and 100% to each pixel, and this can be treated as a lightness value. In other words, a selection is a gray-scale image.

Photoshop's *layer* concept, which I highlighted in Chapter 3 as the concept that, more than any other, led to the product's success, allows you to apply an adjustment to an image in a non-destructive way. A collection of adjustments can be built as a stack of layers, and each layer can be turned on and off, activating and deactivating its adjustment. You can associate with each layer a *mask* that determines which pixels of the underlying image the adjustment is applied to (see Figure E.5). You can probably guess where this is going: the selection can be partial, and the mask is itself … a gray-scale image.

The synergy of these concepts—the fact that channels, masks and selections are all gray-scale images—brings enormous power to the Photoshop user. As an example, suppose you wanted to lighten the skin in a portrait without affecting detail areas such as the eyes, hair, lips, etc. You could open a channel and apply the "find edges" filter to it; this gives a gray-scale image in which the detail pixels are dark. You can now save this image into the copy buffer, and paste it into the mask of a brightness adjustment layer. Don't worry if this seems complicated: it is. But it's what you need to understand to be a Photoshop guru.

Synergistic composition, synchronized states and views. All the compositions I've described can be understood in terms of action synchronizations. For example, the composition of *channel* and *pixel array* in Photoshop synchronizes each edit of a channel with a corresponding edit of the associated pixel array, and vice versa.

But in some of these examples, the net effect of the synchronization might be more easily understood as a synchronization of states. Thus in Photoshop there is an invariant that relates each color pixel to its corresponding gray-scale pixels in the channels; in the RGB case, for example, the lightness value of a pixel in the red channel is exactly the red value of the pixel in the color image.

This suggests an alternative, or additional, form of synchronization, in which concepts are combined by constraints not on their actions but on their states. In this approach, concepts might be identified by projecting the full state of an app into components, and formulating invariants that relate them. In a word processor, for example, the state of a document might be broken into a text component comprising just the characters, and a formatting component that assigns a format to each character independently. More interestingly, a view of the document as a sequence of paragraphs might be separated from a view as a sequence of characters. Inserting a paragraph break adds a new paragraph in the former, but just adds a character in the latter—from an action synchronization perspective, a pinning together of the *add-paragraph* action in one and the *insert-character* action in the other.

This kind of synchronization is related to view structuring, an active area of research in the past which sought to find better ways to modularize systems as overlapping views of a single state in specifications [2, 153] and implementations [109]. In an early paper prefiguring these ideas about concept composition, I used text editing as an example, showing how breaking the specification into two views linked by an invariant makes it possible to write simple specs for actions such as inserting a character and moving the cursor up or down, neither of which could be described easily in one canonical view [61].

72. **Early design of the trash in Windows**. Microsoft introduced the "Recycle Bin" in Windows 95, but, in contrast to Apple's trash, it was not able to store entire folders, and when a folder was deleted, it was disaggregated into individual files. If, however, the user selected for restoring the entire batch of files that had belonged to a folder, the folder was apparently restored with its contents. (See: https://en.wiki-pedia.org/wiki/Trash_(computing))

73. **More composition glitches: when emptying doesn't remove trashed items**. Here's another example of a problem with the trash/folder synergy in macOS. Recall from our discussion of the *unix folder* concept (in Note 15) that an item can have more than one parent folder. Since this is the folder concept used in the Macintosh, it means that a file that has been trashed may remain in another folder. Emptying the trash will remove the file from the trash folder itself, but the file will not have been permanently deleted, and its space will not be reclaimable.

This is an example of an integrity violation (Chapter 11), but it's not especially troublesome because it's not possible (in macOS) to put a file in two folders through the graphical user interface—you need to issue a Unix *ln* command in the console. That said, it is possible that some installed software might come with files linked in two places, and to get rid of them you'd need to track them down in both.

An unwise synergy in Outlook. Attempts at obtaining synergy can also miss the mark. In Outlook, Microsoft's email system, when synchronization between the email client and the message server fails, an error message is written to a log. To store this log and its messages, the designers of Outlook chose to use the existing structure of email folders and messages.

This might have seemed to be a promising synergy between the *log* and *folder* concepts, since it allows the log to be manipulated using the existing tools available for email messages. But the decision created a raft of new problems. Users are surprised to see new mail folders spontaneously appear, and find themselves confused by the large number of messages generated, and frustrated that they can't remove them. Some users have complained that they get into a situation in which their email client attempts to synchronize these error log folders with the server—which fails, of course, because the folders were created as a result of synchronization not working correctly! And system administrators are frustrated because the error logs are generally kept only at the client, so they cannot access them to diagnose problems between the client and server. (See: https://thoughtsofanidle-mind.com/2012/08/29/outlook-sync-issue)

Poor synergies are often examples of a violation of the specificity principle (Chapter 9) in which a concept is *overloaded* with multiple purposes. In the Outlook example, the message *folder* is given a new purpose (storing synchronization logs) that is distinct from, and incompatible with, that of classifying messages.

74. **A Google mystery due to over-synchronization**. I myself was a hapless victim of this strange synchronization flaw. While I was working on this book, I wanted to experiment with Gmail to explore some design issues. So I created a couple of Gmail accounts for fictitious characters named Alice, Bob, and Carol. Sometime later, I noticed that my display name in all my Google apps (in Google Drive, Google Groups, etc.) had switched from Daniel to Alice.

Eventually I figured out what had happened. I had inadvertently created one of the fictitious Gmail accounts in the context of my primary Google account. This had reset my username to "Alice Abalone"—a name that I once thought funny, but since becoming my official name in every professional context in which I used my Google account, was distinctly unamusing.

Amazingly, the change was irreversible. Unable to undo it, I was forced to create a new Google account and switch all my memberships to it. This experience occurred in 2018; hopefully it's been fixed since.

75. **Adobe reverts an update**. This story highlights some of the difficult design trade-offs the Lightroom team faced. While expert users were upset to lose some synchronizations, the developers had reasonably been concerned for less sophisticated users about the extra complexity of the preference settings that controlled the synchronizations.

In a rather remarkable and humble blog post, Tom Hogarty, the head of the Lightroom development team wrote: "We plan to restore the old import experience in our next update … I'd like to personally apologize for the quality of the Lightroom 6.2 release we shipped on Monday. The simplification of the import experience was also handled poorly. Our customers, educators and research team have been clear on this topic: The import experience in Lightroom is daunting. It's a step that every customer must successfully take in order to use the product and overwhelming customers with every option in a single screen was not a tenable path forward. We made decisions on sensible defaults and placed many of the controls behind a settings panel. At the same time we removed some of our very low usage features to further reduce complexity and improve quality."

Somewhat surprisingly, one of the most frequent and anguished complaints was that Adobe had eliminated the synchronization that allowed the flash card or external drive holding the source images to be ejected automatically when the import was complete.

You might have imagined that this was not a major design issue. After all, a user could always eject the card manually in a couple of clicks. It turned out, however, that for many professional users this additional step made the task of uploading photos from a large number of cards feel more burdensome. Complainers also noted that automatically ejecting the card prevents accidental deletions from the card. This entire episode fascinated me: it was a very rare case of a company reverting a change, and an interesting illustration of how sensitive users can be to seemingly small design decisions. Lightroom also happens to be, in my view, one of the very best apps ever.

It should be noted that the update also introduced a bug that was not resolved prior to release; the presence of this bug was also undoubtedly part of the reason for reverting the change. (See: https://blogs.adobe.com/lightroomjournal/2015/10/lightroom-6-2-release-update-and-apology.html)

76. **Under-synchronization in Google Forms**. The lack of synchronization between the visualization and the spreadsheet associated with a Google Form has caught me out a few times. I've often created an anonymous survey, and included at the end an opportunity for responders to add a comment and an email address if they'd like a reply. The submitted email addresses then appear in the summary visualization, compromising their anonymity, but removing them manually from the spreadsheet has no effect. I am thus prevented from sharing the summary with a community that submitted data that was anonymous in every other respect.

77. **Another synchronization issue in Zoom**. The *zoom session* concept allows a single identifier to be used for multiple conversations that occur at different times. The Zoom web portal synchronizes this concept with a conventional *calendar event*, so when you create a session, you are prompted to specify a date and time for it, or a recurring series of dates and times. This synchronization is more complicated than it seems, however. A *zoom session* may have a limit on the number of times it may recur, and will expire if the time between recurrences exceeds some bound. Confusingly, a session scheduled as a single event can recur just like a session scheduled as a recurring event, albeit with a shorter expiry (30 days rather than 365 days). Sessions scheduled as a one-time or recurring event can recur at most 50 times, unless they are scheduled as "recurring events with no fixed time," in which case there is

no limit. My guess is that this synchronization is designed to balance the resource cost of storing session identifiers with the desire to give flexibility to users, but I wonder if it might be accomplished in a simpler way (for example, by listing all unexpired sessions owned by a user in their profile, and encouraging them to delete the ones they do not intend to use again).

78. **Lessons from the Therac-25.** Nancy Leveson and Clark Turner provided a thorough account and analysis of the Therac-25 accidents [92]. Shockingly, the synchronization flaw that I described (and which is explained more fully in the paper) was not discovered after the first accident, which was attributed instead to a hardware glitch; the failure to investigate the accident properly led to multiple additional accidents before it was finally diagnosed correctly.

    Leveson and Turner rightly note that accidents like this happen due to a complex web of missteps, and that it is naive to imagine that software will ever be free of bugs. Nevertheless, the synchronization flaw was indeed the proximate cause of the accidents. The only path to safety (or security) for software-intensive systems runs through simplicity of design, which means using robust concepts whose misfits are well understood (see "Concepts Ensure Safety and Security" in Chapter 3; "Misfits: When Purposes Aren't Fulfilled" in Chapter 5; and Notes 60 and 61).

Chapter 7: Concept Dependence

79. **New concepts must address real problems.** To counter the tendency to complicate a design with unnecessary concepts, each time you are tempted to add a new concept, a good question to ask is: what problem with the existing design does this new concept solve?

    Here are two examples. Netflix added the *profile* concept to its movie-viewing app in 2013, six years after it began online streaming. The problem was straightforward. Different family members watched different movies on the same account, which messed up both recommendations (since my recommendations were based on your prior selections) and placemarks (since my watching a movie cleared the memory that you'd watched only half of it). The *profile* concept nicely solves all these problems by giving each family member essentially their own account— contained in one larger account for billing purposes.

    The Netflix profile concept is really just a second instantiation of the *user* concept in the same app, and thus an example of synergy. Profiles even include a rudimentary form of authentication: the account owner can add a PIN to control access to a profile. This allows the profiles to be used also for parental control of a child's viewing, by limiting a profile to movies with certain ratings. Apparently, customers also use profiles as roles: defining, for example, a "documentary" profile and a "date night" profile to keep movie lists and recommendations separate.

Apple's slide presentation app, Keynote, added the *style* concept about ten years after the first release. Keynote already had a *master* concept that allows you to create master slides that define the styling of a slide's heading, text levels, etc., so a *style* concept might seem unnecessary. Presumably this is why Microsoft PowerPoint still has no *style* concept. In fact, though, the *master* concept is insufficient. It won't let you define a style for things like quotations (which don't occupy a fixed level in a master slide) or indeed any text that appears outside the text body of a master slide; nor does it allow you to maintain consistency across masters (e.g., with a single heading style).

80. **An example of concept as differentiator.** The *identification* concept is what I called a *differentiator* in Chapter 3: a novel concept that the designer hopes will not only be the linchpin of the product but will also differentiate it from competitors. As I noted there, the differentiator may be a technological tour de force: examples are Adobe Photoshop's *layer* concept (which allows non-destructive editing of images), Google Slides's *auto caption* (which generates remarkably accurate captioning on the fly), and WhatsApp's *call* (which gives users free phone calls that require little enough bandwidth that they can run over cellular networks). But differentiators can also be, like our *identification* concept, neither especially subtle nor technologically complex—consider the Tiktok app, for example, whose very successful differentiator is the simple *shared song* concept, which allows users to create videos based on the soundtracks of other users' videos.

81. **Parnas's dependence diagram.** My dependence diagram is inspired by the work of David Parnas, but differs from it in some key respects. The original idea of dependencies was introduced by Parnas in his seminal paper "Designing software for ease of extension and contraction" [116], where he calls it the "uses relation."

The dependencies are *defined* by the code itself, so whether A depends on B is determined by how A is written (and whether, for example, it includes a call to B). Roughly, A depends on B if the correct execution of A relies on the correct execution of B. The consistent subsets that define the product line then *follow* from this definition, because if A relies on B, a subset that includes A but not B will simply not execute properly.

Parnas's proposed strategy is to design the code to have the desired impact in the product line. His methodological principle is this: you should design A to use

B only if you can't imagine a subset that includes A but not B, and furthermore, the use of B should make A easier to build.

Refining Parnas's dependencies. In an early paper of mine, I explained why conventional notions of dependence are not sufficient and may not even make sense [63]. Parnas, to his credit, had foreseen many of these issues, and notes in his original paper, for example, that A calling B is neither necessary nor sufficient for A depending on B. But despite widespread use of the idea of dependencies, these issues have never been fully resolved. More recently, Jimmy Koppel and I proposed a new model of dependence [84] that overcomes some of the problems I raised there (and more), basing the notion of dependence on counterfactual causality.

Concept dependencies vs. code dependencies. Concept dependencies are similar to Parnas dependencies in defining possible subsets, so that the concept dependence diagram (like Parnas's uses relation) characterizes not one application but an entire family.

But concept dependencies differ in a key respect. Concepts are always freestanding and never rely on other concepts for their correct operation. Whereas a code module's dependence is induced by its inherent nature—namely the code inside it, and the calls it makes to other modules—a concept's dependence is a result only of the context of usage.

Concept dependencies are thus more subjective, and embody the designer's assumptions about what makes a consistent app. For the bird song app, maybe *you* thought it was preposterous to even consider an app with questions and answers but without user authentication. In that case, *your* dependence diagram would include not only a dependence of *user* on *q&a*, but also a dependence of *q&a* on *user*, ensuring that any app regarded as consistent include both concepts.

This freedom to use dependencies as a way to express design options comes from the fact that the connections between concepts are expressed in a synchronization when the concepts are composed, and not in the concepts themselves. It also means that removing a concept that is not needed for consistency may involve an adjustment to the synchronizations; for Parnas, in contrast, a module on which a subset does not depend can (theoretically) be simply deleted, and the subset recompiled without it.

In summary: for Parnas, the subsets *follow* from a more basic notion of dependence; for concepts, the subsets *define* dependence.

Does object-orientation violate the dependence principle? Parnas's methodological principle (see above) is often violated. Worse, these violations may be inherent to the way we program, especially in object-oriented code: our most common and familiar idioms seem even to militate *against* Parnas's principle.

Imagine that you're implementing a forum that includes a *post* concept and a *comment* concept, in a language like Java. What would be the standard object-oriented way to implement these concepts? You'd have *Post* and *Comment* classes, with the *Post* class offering methods such as *addComment* and *getComments*. This would induce a dependence (in Parnas's terms) of *Post* on *Comment*.

Now consider the subsets that you'd like your codebase to support. Obviously, it makes sense to have posts without comments, but not to have comments without posts. So the dependence diagram *should* show a dependence of *Comment* on *Post*—just as the concept dependence diagram would show a dependence of the *comment* concept on the *post* concept. So the dependence is the wrong way round, and object-oriented programming seems to have led us to a structure that's upside-down.

How did this happen? The root of this problem is that object-oriented programs are usually structured around control flow. Because the user interface will need to find the comments associated with a post when the post is displayed—and offer a button next to a post to add a comment—it is natural to want to include commenting functionality inside the *Post* class.

In fact, the principles of object-oriented programming make it hard to do anything else. We could give the *Comment* class an internal table that maps posts to comments to support a method like *getComments*, but this time inside *Comment* rather than *Post*, so it would take a post as an argument. Such a table would violate the oft-quoted rule that static state is inimical to object orientation, and classes should not have static components. Another option might be to create a separate class, *Forum* say, that contains tables mapping posts to comments and vice versa. Using such a class would violate other object-oriented principles, however. For example, the fact that a method would need to call a method of *Forum* to obtain a list of comments on a post, and then call their methods to display them, would violate the Law of Demeter [94].

This suggests to me that new programming idioms are needed in which concepts are supported in a more direct and modular fashion in the code. It's not just about dependencies, but also about how the code is organized. Note that in the first object-oriented idiom that I suggested—the naive one that every novice would use, in which *Post* has an *addComment* method—the implementation of the *comment* concept is unfortunately split across classes, since the *Post* class includes the mapping between posts and comments, a structure that belongs to the *comment* concept and not to the *post* concept. A code structure more faithful to the concept design would instead isolate each concept within its own module.

82. **Dependencies emerge from detailed design**. Dependencies rely on quite deep knowledge of the design. In BirdSong 0.1, it's possible that the individual identifications could be upvoted: maybe if an answer said "That's an #american-goldfinch or a #lesser-goldfinch" users could upvote the two identification tags separately. Or maybe users could also upvote recordings just to say they like them. If so, these would produce additional dependencies (of *upvote* on *recording* and *identification* too).

83. **A note about primary and secondary dependencies**. A subset might still form a consistent app if a secondary dependence (a dotted edge) points out of the subset, or if a primary dependence (a solid edge) points out but a secondary dependence is included.

    This reflects the following interpretation of dependencies: each concept depends on *all* of the primary dependees, or on any one of the secondary dependees. A richer notation would let you express fully general dependencies: that a concept depends on a set C of sets of concepts C_i, with the dependence satisfied if whenever the concept is included there is at least a set of concepts that covers one of the C_i. (This corresponds to a disjunctive normal form, and could express any logical combination of dependencies.)

    Feature diagrams. A feature diagram [76] is usually an and/or tree showing the features of a product family, and the combinations that are legal. In terms of specifying combinations, it's comparable to this richer notation (and thus more expressive than my basic diagram). But it doesn't express dependencies directly, so it's less useful from a design point of view.

84. **Facebook's concepts**. The dependence diagram simplifies the concepts of Facebook and their relationships. The *friend* concept has a second purpose, namely to filter which posts a user sees; this is technically an overloading (as explained in Chapter 9).

    In Facebook, you can *tag* a photo to say who's in it (a feature introduced in December 2005), but you can also tag a post or a comment (September 2009). To simplify, I've left the *photo* concept out of the diagram.

    Facebook introduced the *reply* concept in March 2013. Previously, a user could not respond to a comment without adding another comment at the end of the comment list, losing context. Replies introduced threading, so that users could respond directly to a comment. Arguably, *reply* was not a distinct concept from *comment*, but rather an enrichment of the *comment* concept to allow comments not only on posts but also on comments. To represent this in the diagram, we could just give *reply* the new name *threaded comment*, and remove its dependence on *comment*, making it clear that a subset could choose one of the two options: flat comments or threaded comments.

The *like* concept was introduced in February 2009, five years after Facebook's debut. It's now hard to imagine any social media app without an *upvote* or *like* or *reaction* concept, since these concepts are so central to two insidious aspects of the platforms: the psychological stimulus of small rewards that has addictive effect on users, and the value that the platform owner derives from extracting personal information from user preferences.

85. **A paradox in Safari's concepts.** The *private browsing* concept is a bit of a paradox at first sight, but not that puzzling once you understand concept dependencies. It depends on the *cookie*, because if there were no cookies, there would be no need for a concept of private browsing! That is, a subset with *private browsing* but not *cookie* would make no sense. But its very essence is that when private browsing is enabled (in Safari, by opening a private window), cookies aren't used!

Why Safari needs more synergy. My proposed synergy would produce a structure in which *favorite*, *frequently visited* and *reading list* all depend on *bookmark* and are seen as extensions of its functionality, and not independent variants. The *favorite* concept is already largely synergistic: favorites are just a predefined folder within the bookmark collection. The others, however, are largely disjoint. So you can't organize your reading list into folders, for example—that's a bookmark feature, not a reading list feature—and you can't save regular bookmarks for offline reading. As evidence of the confusion that all these related but incomparable concepts produces, I would point to the many articles online that explain the subtle differences between them.

86. **Keynote's concepts.** Keynote actually offers a *character style* concept as well as a *paragraph style* concept. The *shape style* of a shape seems to also store the *paragraph style* of its first paragraph.

The choice of which dependence is primary is sometimes a bit arbitrary. For example, *animation* is given a primary dependence on *special block* because animations are most often used to play the bulleted points in a *special block*. A richer diagrammatic notation that allows multiple dependence sets (without prioritization) could be easily devised but the complexity seems unwarranted (see Note 83).

Two aspects of Keynote's design seem to indicate unresolved design challenges. One is that the *transition* concept, which governs transitions between slides, is distinct from the *animation* concept, leading to some confusion especially when transitions also animate objects (as in the "magic move" transition), and when the presentation is played automatically. The other troublesome aspect revolves around the *special block* concept, which allows paragraphs to be organized in a hierarchy of levels, but is restricted in various ad hoc ways: there can be only one special "body" block on a slide (and regular text blocks cannot have levels), and special blocks can't be grouped. The rationale for these constraints, I believe, is to allow the text in special blocks to be entered in outline mode. In other words, the *special block* concept only makes sense because of the *outline* concept (and thus depends on it).

Chapter 8: Concept Mapping

87. **Dark patterns**. Harry Brignull, a British user experience designer, coined the term "dark patterns" in 2010 for the recurring motifs that websites employ to deceive their users [15]. Most dark patterns involve mapping of concepts to user interfaces. The "Roach Motel" pattern, for example, makes it easy to sign up for something (such as a subscription with a free-trial period) but hard to cancel it.

Usually, the underlying concepts themselves are not implicated. I'm curious to explore, however, whether there are concepts that are themselves designed with ill intent. I have yet to find a convincing example of such a "dark concept" that is used by legitimate companies. But in the security domain, many attacks (such as phishing, cross-site request forgery, injection, and cross-site scripting) might be regarded as dark concepts.

88. **Label mapping in Gmail**. One workaround to the *label* mapping problem in Gmail is simply to switch off the *conversation* view (via a toggle in the preferences). This is not very satisfying, because then you lose the advantage of the conversation structuring.

In its current design, Gmail shows some messages as collapsed and some as expanded. Initially I hoped this might correspond to which messages have the queried label. But this feature seems to be used in multiple, inconsistent ways—including showing the most recent message, showing messages that have been recently modified (e.g., by being starred), and, in the case of the sent message filter, showing messages with that particular label.

Apple Mail has similar issues in the interaction between its *conversation* concept and its *folder* concept. But its conversation setting can be turned on and off on a folder-by-folder basis. By default, conversations are turned *on* for the inbox and *off* for the sent messages folder, so when you ask to see sent messages, that's all you see. This solution would not work in Gmail because it would not generalize to fil-

FIG. E.6 *Trepanning: a metaphor for small design flaws?* (*from Hieronymus Bosch,* The Extraction of the Stone of Madness).

tering on label combinations. (The designers of Gmail have tried to combine the advantages of labels and folders by using label actions to emulate folder actions [129], but this example illustrates one of the limitations of that approach.)

89. **Small design flaws as symptoms of greater pain**. A skeptic might complain that many of the design flaws that I discuss in this book are relatively minor. In response, I would note first that limiting my corpus of examples to major products from leading companies introduces a selection bias, and a larger sample that included products earlier in their development, and products from less capable companies, would reveal larger problems.

I also suspect that many of the design flaws I have discussed, while seemingly small to the onlooker, might have caused considerable pain to the developers who had to deal with the complications they produced in the code. Some of these flaws are thus like trepanning holes, which appear as small defects in exhumed skulls, but are reminders of the pain suffered during their formation (Figure E.6).

90. **Better Backblaze strategies**. There are more efficient ways to search for old files. Instead of looking back one day at a time, you could do a binary search. Say you want to find the latest version of a file before a corruption occurred, and you know the corruption happened sometime between January 1 and March 1. You'd pick February 1 first; then if that version is corrupted, you'd try January 15; if that version is uncorrupted, you'd try February 15 in the hope of finding a later version.

This would reduce your search from 60 to six steps. You might also exploit the file modification date. If the backup of February 1 shows a modification date of January 5, say, there's no point checking January 15; you might as well check January 5 next.

91. **Flags versus labels**. The *flag* and *label* concepts are similar. Flags are usually pre-defined, and mutually exclusive (a maximum of one flag per item). The same filtering conundrum that I describe for flags applies to labels, and indeed, to any concept that involves filtering items by mutable properties.

92. **The live filtering conundrum in Lightroom**. The seemingly obvious mapping for a filtered list of items—in which exactly the items satisfying the filter are displayed—is used in Adobe Lightroom Classic. The usability problems that follow provide a nice case study in the challenges of mapping design.

In Lightroom, a very powerful filtering bar lets you select photos that have been labeled or flagged in various ways, or whose metadata has certain properties, such as including particular keywords, or having been converted to black-and-white. Some of these properties are fixed—for example, the metadata properties that identify the camera or the shooting parameters such as aperture and shutter speed—but most of them can be modified by the user. Adopting the simpler form of mapping, namely maintaining the filter results so they always correspond to the filter setting, even as the items are modified, has some unfortunate consequences.

Here are two examples. Suppose you filter a set of photos to show only those that have been converted to black-and-white. You select one of them from the grid of thumbnails, and, intending to explore different renderings of the image, you open it in "develop mode." If you exit develop mode, you'll return to the thumbnail view, with the photo still selected. The photo being edited is thus always the currently selected image in the collection.

Now, with the photo open in develop mode, you can adjust the tonality, crop the image, and so on. But suppose you select the option to switch the image to color rather than black-and-white. Lightroom permits this adjustment, but now the photo no longer satisfies the criterion for membership in the collection. Not only is the photo instantly removed from the collection (invisibly, since you're still in develop mode), but to maintain the invariant that the photo being edited is always the selected photo in the collection, the develop window suddenly goes blank and displays the message "no photo selected." This is disconcerting and extremely inconvenient: you can't change it back to black-and-white because you're no longer even editing the image! Lightroom's wonderful undo facility can rescue you from this state, although interestingly only by returning you to the thumbnail mode.

That example illustrates a problematic interaction when the user makes a mistake (albeit one that is very easy to make). My second example arises in the context of a task that I often want to perform, and for which, due to this design, I needed to invent a workaround. In Lightroom, you can attach keywords to your images. I do this to identify people in pictures. A common task I want to perform is to check a particular keyword, or perhaps replace it by another one. So let's suppose I keyworded images of my daughter with the keyword "Rebecca" but I want to replace it by the keyword "Becca." So I filter for images carrying the first (old) keyword; to these, I then add the second (new) keyword in bulk. So far so good. Then I remove the old keyword, and all the images disappear, because they no longer carry the keyword of the filter. This is a serious problem, because I want to complete the task by saving the keyword edits in all these photos to their files on disk. Now, without the photos selected, I cannot execute the save command.

(The workaround, incidentally, is to first filter the photos on the required keyword; then to select all; then to turn off the filter, revealing a larger collection in which the relevant photos are still selected. Now they can be edited in bulk without any disappearing as their keywords are changed.)

Following Apple's design of marking the items in the original filtering with their properties would not work here, since the filtering is more sophisticated (allowing multiple properties to be combined). A solution might be to gray out photos that no longer match the filter after they have been edited. The user could then easily see which photos satisfy the filter, but can still select those that do not.

93. **The selection concept and singleton actions**. A similar problem arises with the *selection* concept, which allows a user to select multiple items and then apply an action to them all in aggregate. This is used, for example, in desktop file managers like the macOS Finder, in which you can select multiple files and then delete them all with one click. If there is an action that can only be applied to a single item, invoking that action when more than one item has been selected is problematic.

One solution is simply to block such actions; the Finder does this if you select two files and then try to apply the *rename* action. Another solution is for the selection to mark, in addition to the set of items selected, a single item (within the selected set) that will be the target of such actions. In Lightroom, for example, when you select multiple thumbnails, one of them is highlighted with a slightly brighter frame, and there are user interface actions to cycle through the selection, changing which thumbnail plays this role. This approach could be used to resolve the Lightroom collection-deletion ambiguity problem: when you select multiple col-

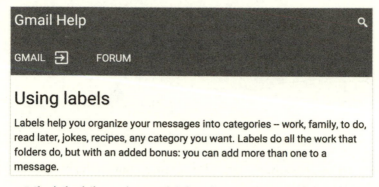

FIG. E.7 *What's the difference between labels and categories? Gmail help isn't too helpful.*

lections, one of them would be distinguished. But this solution seems to me to be overkill in this case, and likely to be confusing to users.

Chapter 9: Concept Specificity

94. **Google's inadvertent humor.** Seeking to understand the difference between labels and categories in Gmail, I looked up "labels" in Google's help documentation. The article began "Labels help you organize your messages into categories ..." (Figure E.7).

95. **Why Zoom has a redundant concept**. Why would Zoom's developers have gone to the trouble of creating an extra concept (*broadcast*) rather than extending an existing one (*chat*)? My hypothesis is that, rather than attempting to integrate the *breakout* concept fully into the app, its engineers opted for a quick-and-dirty solution in which each breakout room is treated as a separate Zoom call. This would explain why "everyone" in the chat of a breakout room refers to only the participants in that room, and why the chat messages of a breakout room are lost when the breakout room is closed.

96. **Apparent redundancy due to distinct purposes.** An apparent redundancy between two concepts may disappear on closer examination, turning out to be a reflection of a genuine difference in purposes. In Adobe Lightroom Classic, the *flag* and *star* concepts appear at first to serve the same purpose: they both let you assign some measure of approval to a photo, and then to filter accordingly.

In fact, however, the two concepts serve different purposes. There are two flag types, "pick" and "rejected," and a dedicated action for deleting all rejected photos. Flags are not saved to the metadata in the photo file itself, so they are intended only

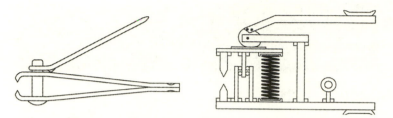

FIG. E.8 *Nail clippers: with function sharing (left) and without (right) (from [145]).*

for temporary use. Stars, on the other hand, range from zero to five, offer actions to increase and decrease, and can be saved to files.

These differences are consistent with the different purposes of the two concepts. The *flag* concept is for selecting and rejecting images as a precursor to deletion; the *star* concept is for rating images that remain, along with their stars, stored over a longer period.

In an earlier version of Lightroom, flags were collection-specific, so you could have separate collections for printing and web display (for example), with different photos flagged in each. I suspect that Lightroom's developers dropped this useful feature because it was confusing to some users.

97. **Concepts in the New Testament**. The Gospel of Matthew seems to have preempted the principle that a concept cannot fulfill two purposes: "No one can serve two masters. Either you will hate the one and love the other, or you will be devoted to the one and despise the other." [Matthew 6:24]

98. **Overloading in mechanical design**. In the design of mechanical systems, it is common to design a single component to have multiple purposes. The sheet metal body of a car not only provides structural support, but also keeps the weather out, and provides an aerodynamic profile. It also acts as the electrical ground for the car.

Karl Uhrich imagined what a nail clipper would look like if no single component was allowed to provide multiple functions, in contrast to the conventional design in which a single metal strip acts both as a spring (through bending) and a cutter (through its sharpened edge) (Figure E.8).

Mechanical designs, in other words, seem to revel in overloading. But software is different. Overloading can be beneficial for mechanical designs because it simplifies the design, saves manufacturing costs, and by reducing size and weight, may lead to better performance. For software, no such considerations apply; two orthogonal concepts are easier to understand than a single concept that serves incompatible purposes, and there is no cost in performance or complexity to a few extra lines of code.

Even in mechanical engineering, overloading can be problematic, and it is advantageous to ensure that different purposes are separated into separately controllable design parameters. *Axiomatic design* [137] is a theory of mechanical design that aims to produce more malleable designs, by ensuring that a change in one functional requirement never requires a change to a design parameter that also influences another functional requirement.

99. **Overloading in social concepts.** Some of the principles of concept design seem to apply not only to software but also to concepts that are executed by people— that is to social structures and policies. I've noticed that overloading by false convergence seems to cause problems particularly in policies related to evaluation and feedback.

Many academic departments (including my own) assign mentors to junior faculty. The mentor is supposed to give encouragement, advice and moral support. When it comes time to consider promotion, the mentor is often one of the first people asked to comment on the case—on the grounds that they have greater knowledge of the candidate's achievements.

By playing these two roles, however, the mentor is put in an untenable position. What if the candidate has revealed inadequacies and concerns that negatively color the mentor's assessment of whether the promotion should go ahead? There is a basic conflict of interest here, which could be eliminated by excluding the mentor from any promotion decisions, or at least instructing the mentor to act, during promotion discussions, only with the interests of the candidate in mind.

In other words, there are two distinct purposes here—providing support and advice on the one hand and helping to assess the candidate on the other—and they are mutually incompatible. Two different roles are needed to serve the two purposes.

The same conflict between assessment and guidance arises in the concept of *review* that conferences use to respond to paper submissions. One purpose of a review is to provide helpful feedback to the author of a paper. A quite different purpose is to decide which papers should appear at the conference. A reviewer cannot satisfy both at once. Even if you rate a paper highly, and would like it to be accepted, if you make constructive suggestions pointing out ways in which the paper might be improved, you run the risk of other committee members using these suggestions as reasons to reject the paper.

To expose such false convergences, you need a clear articulation of the purpose. If the purpose of a submission review is to "review the paper" (merely repeating the name of the concept) or to "solicit expert advice" (which has no value in itself), the issue won't arise. You need to articulate honest and straightforward purposes such as "choose which papers to accept" and "give helpful feedback to authors" to make these conflicts apparent.

Another example of overloading in social settings is provided by Kieran Egan whose pedagogical theory [37] starts from the observation that the three conventional purposes of education—socializing students to prevailing norms, teaching them to seek higher truths and transcend prejudices, and helping them fulfill their own personal potential—are fundamentally incompatible. Egan insinuates, more generally, that the success of any social institution may rest on the degree to which its purposes are aligned. Prisons, for example, are problematic because their purposes—punishment and rehabilitation—are in direct opposition, and achieving one can only come at the expense of the other.

100. **More on Epson's overloading**. You might think that the overloading in Epson's printer driver that binds the *paper feed* setting to the *paper size* is mitigated by the ability to choose the paper feed in the printer dialog itself.

This is in fact possible, but the Epson driver won't let you pick a value that is incompatible with the value in the paper size setting. Needless to say, this confuses many users, who don't realize that their paper size selection determined the feed choice and are then puzzled that their preferred feed option is grayed out in the dialog. The overloading flaw is thus exacerbated by Epson's excessive conservatism in saving the user from errors (explained in the *Over Synchronization* section of Chapter 6).

101. **Overloading in Photoshop's cropping function**. For photography aficionados, here's a richer example of overloading that was eventually fixed by Adobe.

Adobe Photoshop has a *cropping* concept. Its purpose is to allow you to trim the edges off an image so you can remove unwanted parts of a photograph. The operational principle is: you create a cropping frame within the image whose dimensions and position you can adjust; then if you issue the crop command, the pixels outside the frame are removed.

Or at least that is the operational principle in the current version of the app. Until a few years ago, cropping in Photoshop was much more complicated than this. Let's look in particular at one part of the user interface (Figure E.9).

FIG. E.9 *Cropping in Photoshop CS5.*

Notice that there are fields in which you can enter a width and height for the cropping frame. This supports an important feature: the ability to maintain a fixed aspect ratio. If you want to print on paper of a certain shape, or you want to maintain consistency between the images in a portfolio, you might want your image to conform to a particular aspect ratio—such as 2 × 3.

By using the width and height fields, you can set a fixed aspect ratio. As you now adjust the cropping frame, the aspect ratio will be maintained. If you look carefully at these fields, however, you'll notice that the values that are entered include not only the numbers but also the *units*. In this example, the units are set to inches, so what I've actually specified is not 6 × 4 but *6 in × 4 in*.

Cropping now has two, independent consequences. One is that you remove the pixels outside the cropping frame. The other is that the new image will be resized to have the given dimensions. If the photo was not *6 in × 4 in* before, it will be afterwards. That will affect the default printing size in many applications.

But wait, there's yet more complexity! If a resolution is specified in the dialog, the cropping action will preserve the resolution of the file. Since the dimensions in inches are being changed, and the resolution is fixed, that means that there will have to be a change in the number of pixels, by either upsampling or downsampling the image.

Suppose the image were $6\,in \times 4\,in$ to start with, and 200 pixels per inch. Then the image had $800 \times 1200 = 960{,}000$ pixels. If I crop away half of the image, half the original pixels will be removed. But if both the dimensions and the resolution are to be maintained, the overall pixel count cannot change, so the image will need to be resampled.

One rather surprising consequence of this is that if you define the cropping frame to include the entire image—that is, with no pixels outside the frame—the cropping action may still modify the file!

The problem here is that there is a separate concept of *resampling* which has been piggybacked onto the *cropping* concept. The result is not only a confusingly complicated interface for non-experts, but also the strange inability to set a fixed aspect ratio without modifying the image dimensions.

This design flaw was fixed in Adobe Photoshop CS6 (2012) by teasing apart the two concepts. The features of the *resampling* concept no longer appear in the dialog for *cropping*, and there is a new aspect ratio option that allows you to set the aspect ratio in dimensionless units.

102. **Recommendation, upvote, and karma concepts**. In my analysis of the Facebook *like* concept, I identified the *recommendation* concept and not the *upvote* concept as serving the purpose of curating your newsfeed. The *upvote* concept aggregates approvals and disapprovals of an item to rank that particular item; this is how newspapers sort reader comments, and forums such as Reddit and Hacker News highlight the most popular posts. The *recommendation* concept, in contrast, uses approvals of items in the past to predict the relevance of items in the future. It is related to the concept of *karma*, in which posts in a public forum are ranked according to the reputation of the contributor, measured by aggregating ratings of their prior contributions.

Would my Facebook solution really work? I suggested that the Facebook *like* concept be split up so that the user has separate control over sending reactions and curating their feed. In Facebook's case, I actually doubt this would help, because the curation algorithm is so opaque that Facebook users don't have any understanding of how posts are selected and ordered in their newsfeed, so they would be unlikely to take the extra trouble of providing separate information for that purpose. Perhaps in a new platform that is more committed to the interests of the user, and less driven by advertising (if such a thing were possible), this division would be useful. I will note also in its defense that it could be mapped to two clicks—one to select the emotional reaction and one for thumbs up or down—which is no more than the Facebook design currently takes.

A design problem with the upvote concept. The *upvote* concept suffers from an overloading problem in many cases, since it is often used to *signal* support or antagonism. This purpose is in conflict with the purpose of obtaining accurate measures of approval, but I don't know of any concept split to resolve it, assuming that the vote tally must be public. The problem, by the way, is related to a problem that arises in student evaluations of teachers, in which a small number of angry students submit an artificially low rating in order to bring down a teacher's overall average. In my own experience, such cases can be spotted in the response data because such students give the lowest possible rating to *every* aspect of the teacher and the course, and in their informal comments frequently reveal that they felt unfairly graded.

Chapter 10: Concept Familiarity

103. **Normal and radical design**. In *What Engineers Know and How They Know It* [147], Walter Vincenti distinguishes between "normal" and "radical" design. Normal design involves the refinement and extension of an accepted, standard design. A car designer, for example, takes the essential structure for granted: that the car will have four wheels, a conventional gasoline or hybrid engine, a gearbox, etc. Because normal design uses familiar components in familiar ways, and relies on a large body of experience, the designer can be confident that the design will perform as expected.

Radical design is much rarer. It's what NASA's engineers did when they designed the lunar module of Apollo 11. Nobody had ever designed such a thing before, and they could not be sure that it would work at all. In fact, as the lunar module descended, it generated alarms indicating that the computer was overloaded and had to shed some tasks (fortunately not including the ones that were needed to ensure a safe landing).

In software, the obviously normal designs are the countless "CRUD" applications being built using content-management platforms such as Drupal and WordPress. (Of course, like all normal designs, these were at one time radical.) Radical designs include the first graphical user interfaces (invented at Xerox PARC and then realized commercially by Apple), the first spreadsheet (Dan Bricklin's VisiCalc), and the first relational databases (based on Edgar Codd's design).

In practice, the distinction is not binary, and most designs—including software applications—lie between the two extremes. Sometimes a design is radical in its scope and overall purpose, despite the reuse of normal concepts—Tim Berners Lee's invention of the World Wide Web, for example, which exploited decades-old concepts (notably *hypertext* and *markup*) to dramatic new effect. Sometimes a design is not radical in its purpose but introduces concepts that change the game entirely; Adobe Photoshop did this with the introduction of its *layer* and *mask* concepts.

Not all novelty in a product is the result of radical design. In software, new concepts emerge over time, in response to new uses and new opportunities. Twitter's *hashtag* concept solidified and extended an idea that users had invented. This follows a pattern that Eric von Hippel [56] has argued is typical in many industries: that innovation often comes not from the suppliers or products, but from their users, who first feel the need for new functionality, and often even create it in prototype for themselves.

The key point is that radical design is atypical. Normal design is what almost all designers do day to day. This doesn't mean designers just reinvent the wheel. Even in a normal design, there are many opportunities for small innovations, resulting over time in dramatic technological advances. It also doesn't mean that design does not matter. On the contrary, the quality of a normal design can spell the difference between success and failure, for any product, including software.

To be radical, a design does not need to change everywhere. Changing one critical component can make a design radical, while keeping others fixed to reduce risk. The designers of the first hybrid car switched the engine but they didn't replace the steering wheel with a joystick, or use acrylic in place of glass for the windows, or alter the shape of the cabin.

In software, similarly, one new concept can make a design radical. The radical move in the design of the World Wide Web—in my view the key insight of its inventor—was, as I argued in Chapter 3, the concept of the *url*, whose purpose is to provide a distinct and persistent name for each resource that is independent of where and how the resource is stored. This was what made the Web different from all its predecessors (such as HyperCard), making possible a massive shared information infrastructure.

On a smaller scale, Adobe Lightroom's designers based image editing not on Photoshop's *layer* and *mask* concepts, but on the concept of an *action*—an image adjustment that is stored in the image's metadata as part of a history, making possible flexible and non-destructive image editing. The implications of this change are enormous. Interaction becomes simpler; non-destructive edits become easy; modifications can be stored as metadata in the image files themselves; and the file sizes are dramatically reduced.

104. **Alexander's design patterns**. The idea of capturing design expertise in generic "design patterns" that can be instantiated in different contexts originated in the work of the architect Christopher Alexander [4, 5]. Alexander is known to computer scientists through the "Gang of Four," whose influential collection of design patterns in object-oriented programming [44] cited him as an inspiration.

Concepts are in fact closer to Alexander's design patterns than the Gang of Four patterns are, because, like Alexander's patterns, concepts are driven by the needs of users, and shape users' experiences of the product; the Gang of Four patterns, on the other hand, are motivated mostly by the needs of programmers (in particular making code easier to evolve over time). At heart, however, design patterns and concepts have a lot in common, and the Gang of Four deserve credit for changing the way we think about programming.

For Alexander, patterns address the fundamental challenge of design, namely the unknowability of context that results in unanticipated misfits (see Note 60). Because patterns represent normalized design (see Note 103), they embody long experience in coping with the misfits that typically arise, and save you from the design mistakes you would make if you were designing from scratch.

Computer scientists who have encountered Alexander only secondhand through the literature on design patterns in software may be surprised to discover, on reading Alexander's work, that he writes not so much as a designer, let alone as an engineer, but more as a poet, reveling in the spirituality of good design (especially in his latest books [6]). The spiritual and aesthetic components of concept design have yet to be explored, but may be essential if we want to go beyond software that works to software that delights and inspires.

105. **Why does PowerPoint have a cursor?** In my discussion of PowerPoint's *section* concept, I referred to "the selected slide." Even behind this simple phrase lurks a conceptual design question. Many apps involve manipulating a sequence of items by insertion, deletion, and moving. All of them have some concept of *selection*; what distinguishes them is whether you can select more than one item, and if so, whether those items must be contiguous.

Many also have a concept of *cursor*, which marks the point at which an insertion will occur. In text editors, the relationship between the cursor position and the current selection is quite complicated; in the editor I'm typing this text in (BBEdit), for example, if I select a word, the cursor disappears; advancing the cursor (with the right arrow key) then places it after the word, and moving it back (with the left arrow key) places it before the word.

One reason that text editors have a cursor in addition to selections is to support replacements: if you make a selection, and then start typing, the new characters *replace* the characters that were selected. Placing the cursor without a selection allows you to insert without making any replacements.

This complication—of having both a cursor and allowing selections—works well in text editors but doesn't seem necessary in a slide presentation app. So Keynote has no cursor; if you select a slide and add a new slide, the new one just appears after the selected slide (without replacing it, of course). But both PowerPoint and Google slides include both cursor and selections. Unlike in a text editor, where a simple click always sets the cursor, in these apps you have to click quite carefully *between* slides to place the cursor without selecting a slide.

I'm not sure why this complication is useful. You might think that it makes adding slides a little more intuitive, because you don't need to remember whether the new slide comes before or after the selected one. So to add a new slide at the beginning of a section in PowerPoint, you might place the cursor just before that slide. Sadly, you can't do that though: the cursor can only be placed between slides within a section or after the last slide.

Perhaps this whole discussion seems nitpicky. But while any one small complexity like this might not matter, the accumulation of many such needless complexities exacts a heavy price, from programmer and user alike.

106. **Inevitability as a design principle**. The PowerPoint example illustrates a general design principle. Consider how the *add section* command might have behaved. There are so many different possibilities: to include or exclude the subsequent slides from the newly created section; if a slide is currently selected, to include or exclude that particular slide; to allow the action to be executed when non-contiguous slides are selected or not; and so on.

Whenever a design decision is made from such a set of possibilities, and when those possibilities seem equally plausible—or at least none stands out as obviously better than the others—the designer runs the risk of making a non-optimal choice. So the very presence of all those options leads to a fragile design process, in which at every step the designer is likely to make a mistake, following a path along which arbitrary decisions take the designer towards a more idiosyncratic and incoherent design. Each of these choice points is faced by the poor user too, who has to guess how the design will behave from amongst the equally plausible options.

These decision points are thus a symptom of a bad design. In a good design, the design decisions seem to be *inevitable*. Only one of the options is plausible, and if more than one is plausible, the choice is resolved by following a general rule or sensibility that applies uniformly to all decisions within the product. The user will be able to predict which behavior is most likely at any point, either because only one makes sense, or because the user will be aware (perhaps only subconsciously) of, and guided by, the general rules and sensibilities that have been conveyed by the design and the detailed behaviors the user has encountered so far.

For the designer, this raises a question. When you reach a decision point at which there seem to be many possibilities, what do you do? You should start, of course, by evaluating the options. If one is clearly and demonstrably superior to the others—which means that in a team you will have consensus—you should probably choose that option. If not, you can try to formulate a general principle that would favor one over the others. If you're able to do that, and can apply that principle throughout your design, choosing that one option may be justified. If you can't do that, you have no choice in my view but to back off: to undo the previous decision that brought you to this point of confusion.

The inevitability of your design move at each point, in other words, is not only a mark of design quality, but evidence that the decisions you have made so far have been sound.

107. **More on Lightroom's unconventional export presets**. The unusual semantics of the augmented *preset* concept is reflected also in its mapping to the user interface. In the standard checkbox widget, the label that sits next to the box is not usually selectable as a separate control. This is what Don Norman would call an affordance problem: the preset names don't signify that they have a clicking affordance distinct from toggling the checkbox.

Complex concepts confuse technical writers. The complexity of the *preset* concept is evident in the documentation too. The help page includes an FAQ with the question "Why are some sections hidden when presets are checked?" The answer given is as follows: "When you select one or more export presets in the Export dialog, Post-Processing section and other sections created by third-party plug-in are hidden in the Export Settings. However, the export settings defined for Post Processing and other sections from third-party plug-ins in the export preset are respected and images are exported accordingly." Hmm.

Complex concepts confuse programmers. While experimenting with the preset dialog in order to understand it better, I found that the entire application sometimes became unresponsive and I had to force quit it and restart. It seems likely to me that the conceptual complexity of the design has resulted in some untamed complexity in the code, and Adobe may be experiencing some trepanning here (Note 89).

108. **Using nicknames in contacts**. As evidence that many people use nicknames and not real names for their contacts, consider the fact that NokNok, an Israeli start-up, actually marketed an app to exploit this. Users who downloaded the app were connected and shown the nicknames others used for them in their contacts. Perhaps not surprisingly, the company eventually switched business areas and is now focused on providing free VOIP calls.

Apple's Contacts app in fact supports the *nickname* concept. You can select "nickname" from a set of fields that can be added to a contact. What you enter there appears to be private: if you start typing a nickname in an email message, the address is completed but the nickname does not appear in the message that is sent. With this feature, our Prince can safely call the Queen "Mummy" after all.

Chapter 11: Concept Integrity

109. **Robustness of mental models**. In early and influential work on mental models, Johan de Kleer and John Seely Brown argued that only certain kinds of models served users well, allowing them to reliably predict behavior (especially in novel situations). These models, which they termed "robust," had to satisfy certain "aesthetic principles" [80].

Their first and most important principle, which they called "no-function-in-structure," required that the behavior of the components of a system be context free. This would mean, for example, that the explanation of how a switch works could not refer to the function of other parts of the circuit (even though of course whether a component is activated may depend on whether the switch is on or off).

It is reassuring that this principle, emerging from a psychological perspective, aligns so closely with the idea of concepts as independently explainable units of functionality, and with the integrity principle (see "Concepts are freestanding" in Note 48).

110. **Feature interactions and integrity**. Concept integrity is strongly related to the idea of feature interaction in telephony (see "How concepts differ from features" in Note 48).

One particular formulation—which appears as the definition of feature interaction in [122], and as the last of a series of possible definitions in [22]—associates a system-level specification with each feature, and then says that an interaction exists if the presence of one feature causes the specification of another to no longer hold.

This is exactly the definition of concept integrity. Ruling out such interactions is, by the standards of the feature interaction literature, quite extreme. But it seems essential to preserving the nature of concepts; without it, concepts would not be understandable in their own right, because their behavior would depend on the context in which they are deployed.

In practice, concept integrity does not seem to impose unreasonable demands, for several reasons. First, not all features need be implemented as concepts. Classic telephony features such as "call forward on busy" and "voice mail on busy" would probably be better represented as synchronizations of concepts than as concepts in their own right. Second, distinct features in telephony (such as "selective call acceptance" and "selective call rejection") may be contained within a single concept (in this case one that manages accepting and rejecting calls based on preference lists).

Third, a concept specification only governs its own actions, so when concepts are synchronized in collaborative composition, one concept need not impinge on the behavior of another. For example, if call forwarding is described in terms of mapping phone numbers to phone lines, and regular calling involves only making connections between lines (and has nothing to say about numbers), the two can be composed without risk of integrity violation.

Finally, when concepts are composed, the mapping to a user interface can change which actions are exposed to the user. For example, an *email* concept might have a *delete* action for permanently deleting messages. When composed with the *trash* concept, however, the deletion button in the user interface will no longer be associated with the *email.delete* action, but rather with *trash.delete*, the deletion action of the *trash* concept. In such a case, it might appear to the user that integrity is violated—deletion no longer permanently removes a message. But once the user understands the mapping of user interface controls to concept actions, integrity is recovered.

111. **Font magic in Apple Pages and other delights of format-toggle.** In Apple Pages, if you bold some text in Helvetica Light, it will be in Helvetica Bold; bolding again takes you back to Helvetica Light. And if you start in Helvetica Regular, and apply bold twice, it will take you back to Helvetica Regular. This sounds nice, because it preserves the toggling, until you realize that the text that was supposedly in Helvetica Bold is treated differently in the two cases! The app is apparently remembering more about the format setting than is evident in the formatting dialog, and so integrity is violated for other reasons.

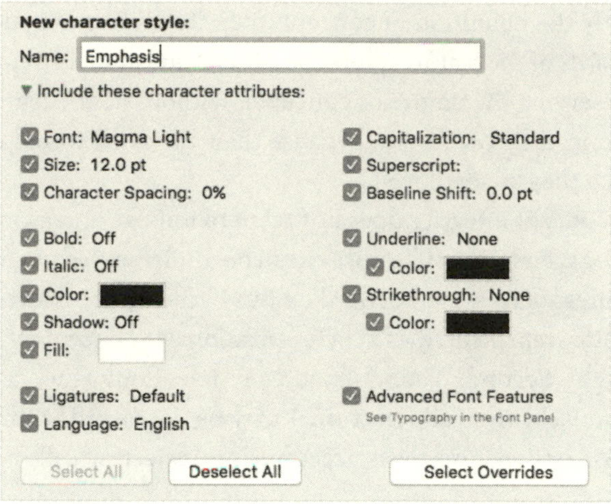

FIG. E.10 *An attempted solution to the formatToggle integrity problem, in Apple Pages '09.*

(You can reveal that something funny is going on as follows. Start with some text in Helvetica Light, say. Now bold it. If you bold again, you'll be back to Helvetica Light. But if before you do that, you pull down the font style menu showing Bold, and you click on that highlighted menu item, you will now have cleared the hidden state, and it will revert to Regular when bolded again.)

It's important to note that Apple's style mechanism does *not* permit partial styles to be defined, losing the key benefit of separating out bold and italic as settings independent of the font subfamily. The 2009 version of Apple's productivity apps *did* allow partial styles (see Figure 8.12 in Chapter 8), but this had other problems. Instead of treating a professional font as a single typeface family, it broke the family down into "subfamilies"—a classification defined for both TrueType and OpenType fonts. In the screenshot of Apple Pages '09 (Figure E.10), you can see that the "font" is set to "Magma Light," which represents one of the subfamilies of the Magma typeface. If the subfamily has exactly the four variants (regular, bold, italic, bold italic), this works nicely. But some subfamilies (like Magma Light) are defined by their weight, so they don't have a bold variant!

Adobe InDesign does have the *format toggle* concept insofar as having bold and italic actions. These suffer from the same problem as TextEdit's. But unlike in Apple's productivity apps, bold and italic are not available as settings for styles, so are missing where they would be most useful.

112. **No backup in Google drive.** Google Drive has no built-in backup facility. It does save old versions of files—a valuable feature—but if the file itself is deleted, the old

versions are deleted along with it. Google's synchronization utility is confusingly called "Backup and Sync," but the backup refers to backing up files from your computer to the Google Drive, not to backing up in the other direction.

113. **More on the Google Drive accident**. When Google Drive synchronizes a folder that is empty on your local machine, it doesn't permanently delete the files in the cloud, but instead moves them to the trash. Unfortunately, in the scenario I described, the user emptied the trash: "I was organizing my files on my local computer. I moved them around and out of my Google Drive folder which syncs files. I didn't think anything of it. In the process I got an email from Google saying I'm running out of storage. So I go to the Google Drive site and empty the trash. I didn't think anything of it."

The trash is hardly a protection for the problem described here, because an unsuspecting user might attempt to move files out of a synchronized folder *precisely* to allow the trash to be emptied and thus to gain space in their drive.

For the full sad story, see the web page http://googledrivesucks.com. Perhaps one measure of the seriousness of a usability problem is whether it causes enough pain for someone to go to the trouble of registering a domain in protest.

114. **This book's website and forum**. To spare you the trouble of registering your own domain to complain about this book, I have created a website that includes a link to a discussion forum for topics related to the book and to concept design. Please visit https://essenceofsoftware.com. I look forward to your participation.

תושלב״ע

References

[1] Jean-Raymond Abrial. *The B-Book: Assigning Programs to Meanings*. Cambridge University Press, 2005.

[2] M. Ainsworth, A. H. Cruikchank, P. J. L. Wallis, and L. J. Groves. Viewpoint specification and Z. *Information and Software Technology*, 36(1):43–51, 1994.

[3] Christopher Alexander. *Notes on the Synthesis of Form*. Harvard University Press, 1964.

[4] Christopher Alexander. *A Pattern Language: Towns, Buildings, Construction*. Oxford University Press, 1977.

[5] Christopher Alexander. *Timeless Way of Building*. Oxford University Press, 1979.

[6] Christopher Alexander. *The Nature of Order: An Essay on the Art of Building and the Nature of the Universe* (4 volumes). Center for Environmental Structure, 2002.

[7] Charles Bachman and Manilal Daya. The Role Concept in Data Models. *Proceedings of the Third International Conference on Very Large Data Bases*, Tokyo, Japan, Oct. 6–8, 1977, pp. 464–476.

[8] Don Batory and Sean O'Malley. The design and implementation of hierarchical software systems with reusable components. *ACM Transactions on Software Engineering and Methodology*, Vol. 1:4, Oct. 1992, pp. 355–398.

[9] Nels E. Beckman, Duri Kim, and Jonathan Aldrich. An empirical study of object protocols in the wild. *Proceedings of the European Conference on Object-Oriented Programming* (ECOOP '11), 2011.

[10] Dines Bjørner. Software systems engineering—From domain analysis via requirements capture to software architectures. *Asia-Pacific Software Engineering Conference*, 1995.

[11] Dines Bjørner. *Domain Engineering: Technology Management, Research and Engineering*. Japan Advanced Institute of Science and Technology (JAIST) Press, March 2009.

[12] Gerrit A. Blaauw and Frederick P. Brooks. *Computer Architecture: Concepts and Evolution*. Addison-Wesley Professional, 1997.

[13] Laurent Bossavit. *The Leprechauns of Software Engineering: How Folklore Turns into Fact and What to Do about It*, 2017.

[14] Douglas Bowman. *Goodbye, Google*. 20 March, 2009. At https://stopdesign.com/archive/2009/03/20/goodbye-google.html.

[15] Harry Brignull. Dark Patterns. At https://www.darkpatterns.org.

[16] Robert Bringhurst. *The Elements of Typographic Style*. Hartley & Marks, 1992

[17] Frederick P. Brooks. *The Mythical Man-Month*. Addison-Wesley, Reading, Mass., 1975; Anniversary edition, 1995.

[18] Frederick P. Brooks. No silver bullet—essence and accident in software engineering. *Proceedings of the IFIP Tenth World Computing Conference*, 1986, pp. 1069–1076.

[19] Frederick P. Brooks. *The Design of Design: Essays from a Computer Scientist*. Addison-Wesley Professional, 2010.

[20] Julien Brunel, David Chemouil, Alcino Cunha and Nuno Macedo. The Electrum Analyzer: Model checking relational first-order temporal specifications. *Proceedings of the 33rd ACM/IEEE International Conference on Automated Software Engineering* (ASE 2018), Association for Computing Machinery, New York, NY, USA, pp. 884–887, 2018.

[21] Jerome Bruner. *Toward a Theory of Instruction*. Harvard University Belknap Press, 1974.

[22] Glenn Bruns. Foundations for features. In S. Reiff-Marganiec and M.D. Ryans (eds.), *Feature Interactions in Telecommunications and Software Systems VIII*, IOS Press, 2005.

[23] Jerry R. Burch, Edmund M. Clarke, Kenneth L. McMillan, David L. Dill and L. J. Hwang. Symbolic model checking: 10^{20} states and beyond. *Information & Computation* 98(2): 142–170, 1992.

[24] William Buxton. Lexical and pragmatic considerations of input structures. *ACM SIGGRAPH Computer Graphics*, Vol. 17:1, January 1983.

[25] Stuart Card and Thomas Moran. User technology—From pointing to pondering. *Proceedings of The ACM Conference on The History of Personal Workstations* (HPW '86), 1986, pp. 183–198.

[26] Stuart K. Card, Thomas P. Moran, and Allen Newell. *The Psychology of Human-Computer Interaction*, Lawrence Erlbaum Associates, 1986.

[27] Peter Chen. The entity-relationship model—Toward a unified view of data. *ACM Transactions on Database Systems*, Vol. 1:1, March 1976, pp. 9–36.

[28] Michael Coblenz, Jonathan Aldrich, Brad A. Myers, and Joshua Sunshine. Interdisciplinary programming language design. *Proceedings of the ACM SIGPLAN International Symposium on New Ideas, New Paradigms, and Reflections on Programming & Software* (Onward! 2018), 2018.

[29] Richard Cook and Michael O'Connor. Thinking about accidents and systems. In K. Thompson and H. Manasse (eds.), *Improving Medication Safety*, American Society of Health-System Pharmacists, 2005.

[30] Nigel Cross. *Design Thinking: Understanding How Designers Think and Work*, Bloomsbury Academic, 2011.

[31] David L. Detlefs, K. Rustan M. Leino and Greg Nelson. Wrestling with rep exposure, SRC Report 156, Digital Systems Research Center, July 29, 1998.

[32] Edsger W. Dijkstra. The Structure of the "THE"–multiprogramming system. *ACM Symposium on Operating System Principles*, Gatlinburg, Tennessee, October 1–4, 1967.

[33] Edsger W. Dijkstra. A position paper on software reliability (EWD 627). 1977. At http://www.cs.utexas.edu/users/EWD/transcriptions/EWD06xx/EWD627.html.

[34] Edsger W. Dijkstra. On the role of scientific thought (EWD 447). 1974. At http://www.cs.utexas.edu/users/EWD/ewd04xx/EWD447.PDF. Also in: Edsger W. Dijkstra, *Selected Writings on Computing: A Personal Perspective*, Springer-Verlag, 1982, pp. 60–66.

[35] Edsger W. Dijkstra. On anthropomorphism in science (EWD936), 25 September 1985. At https://www.cs.utexas.edu/users/EWD/transcriptions/EWD09xx/EWD936.html.

[36] Edsger W. Dijkstra. The tide, not the waves. In *Beyond Calculation: The Next Fifty Years of Computing*, Peter J. Denning and Robert M. Metcalfe (eds.), Copernicus (Springer-Verlag), 1997, pp. 59–64.

[37] Kieran Egan. *The Educated Mind: How Cognitive Tools Shape Our Understanding*. The University of Chicago Press, 1997.

[38] Eric Evans. *Domain-Driven Design: Tackling Complexity in the Heart of Software*. Addison-Wesley, 2004.

[39] Rolf A. Faste. Perceiving needs. *SAE Journal*, Society of Automotive Engineers, 1987.

[40] Robert W. Floyd. Assigning meanings to programs. *Proceedings of Symposia in Applied Mathematics*, Vol. 19, 1967, pp. 19–32.

[41] James D. Foley and Andries van Dam. *Fundamentals of Interactive Computer Graphics*. Addison-Wesley Publishing Company, 1982.

[42] Martin Fowler. *Analysis Patterns: Reusable Object Models*. Addison-Wesley Professional, 1997.

[43] Richard Gabriel. Designed as designer. *23rd ACM SIGPLAN Conference on Object-Oriented Programming, Systems, Languages and Applications* (OOPSLA '08), Oct. 2008.

[44] Erich Gamma, Richard Helm, Ralph Johnson, and John Vlissides. *Design Patterns: Elements of Reusable Object-Oriented Software*. Addison-Wesley Professional, 1994.

[45] Joseph A. Goguen and Malcolm Grant (eds.). *Software Engineering with OBJ*. Springer, 2000.

[46] Thomas R. G. Green. Cognitive dimensions of notations. In A. Sutcliffe and L. Macaulay (eds.), *People and Computers*. Cambridge University Press, pp. 443–460, 1989.

[47] Thomas R. G. Green and Marian Petre. Usability analysis of visual programming environments: a 'cognitive dimensions' framework. *Journal of Visual Languages & Computing*, June 1996.

[48] Saul Greenberg and Bill Buxton. Usability evaluation considered harmful (some of the time). *Proceedings of Computer Human Interaction* (CHI 2008), Apr. 2008.

[49] Carl A. Gunter, Elsa L. Gunter, Michael Jackson and Pamela Zave. A reference model for requirements and specifications. *IEEE Software*, Vol. 17:3, May 2000, pp. 37–43.

[50] John Guttag and J. J. Horning. Formal specification as a design tool. *Proceedings of the 7th ACM SIGPLAN-SIGACT Symposium on Principles of Programming Languages* (POPL '80), 1980, pp. 251–261.

[51] John V. Guttag and James J. Horning. *Larch: Languages and Tools for Formal Specification*. Springer, 1993 (reprinted 2011).

[52] Michael Hammer and Dennis McLeod. Database description with SDM: A semantic database model. *ACM Transactions on Database Systems*, Vol. 6:3, Sept. 1981, pp. 351–386.

[53] David Harel. Dynamic logic. In Gabbay and Guenthner (eds.), *Handbook of Philosophical Logic*. Volume II: Extensions of Classical Logic, Reidel, 1984, p. 497–604.

[54] Michael Harrison and Harold Thimbleby (eds.). *Formal Methods in Human-Computer Interaction*. Cambridge University Press, 2009.

[55] Ian Hayes (ed.). *Specification Case Studies*. Prentice Hall International, 1987.

[56] Eric von Hippel. *Free Innovation*. MIT Press, 2017. Full text at https://papers.ssrn.com/sol3/papers.cfm?abstract_id=2866571.

[57] C.A.R. Hoare. The emperor's old clothes. *Communications of the ACM*, Vol. 24:2, 1981, pp. 75–83.

[58] C.A.R. Hoare. *Communicating Sequential Processes*. Prentice-Hall, 1985.

[59] Walter Isaacson. *Steve Jobs*. Simon & Schuster, 2011.

[60] Edwin Hutchins, James Hollan and Donald Norman. Direct Manipulation Interfaces. *Human-computer Interaction*. Vol. 1:4, Dec. 1985, pp. 311–338.

[61] Daniel Jackson. Structuring Z specifications with views. *ACM Transactions on Software Engineering and Methodology*, Vol. 4:4, 1995, pp. 365–389.

[62] Daniel Jackson and Craig A. Damon. Elements of style: analyzing a software design feature with a counterexample detector. *IEEE Transactions on Software Engineering*, Vol. 22:7, July 1996, pp. 484–495.

[63] Daniel Jackson. Module dependencies in software design. *9th International Workshop on Radical Innovations of Software and Systems Engineering in the Future* (RISSEF 2002), Venice, Italy, Oct. 2002, pp.198–203.

[64] Daniel Jackson, Martyn Thomas, and Lynnette Millett, eds. *Software for Dependable Systems: Sufficient Evidence?* National Research Council. National Academies Press, 2007. http://books.nap.edu/openbook.php?isbn=0309103940.

[65] Daniel Jackson. A direct path to dependable software. *Communications of the Association for Computing Machinery*, Vol. 52:4, Apr. 2009, pp. 78–88.

[66] Daniel Jackson. *Software Abstractions*. MIT Press, 2012.

[67] Daniel Jackson. Alloy: A language and tool for exploring software designs. *Communications of the ACM*, Vol. 62:9, Sept. 2019, pp. 66–76. At https://cacm.acm.org/magazines/2019/9/238969-alloy.

[68] Michael Jackson. *System Development*. Prentice Hall, 1983.

[69] Michael Jackson. *Software Requirements and Specifications: A Lexicon of Practice, Principles and Prejudices*. Addison-Wesley, 1995.

[70] Michael Jackson. *Problem Frames: Analysing & Structuring Software Development Problems*. Addison-Wesley Professional, 2000.

[71] Michael Jackson. *The World and the Machine*. At https://www.theworldandthemachine.com.

[72] Michael Jackson. The operational principle and problem frames. In Cliff B. Jones, A. W. Roscoe and Kenneth R. Wood (eds.), *Reflections on the Work of C.A.R. Hoare*, Springer Verlag, London, 2010.

[73] Ivar Jacobson. *Object Oriented Software Engineering: A Use Case Driven Approach*. Addison-Wesley Professional, 1992.

[74] Natasha Jen. *Design Thinking Is Bullsh*t*. 99U Conference, 2017. Video online at: https://99u.adobe.com/videos/55967/natasha-jen-design-thinking-is-bullshit

[75] Cliff B. Jones. *Systematic Software Development Using VDM*. Prentice Hall, 1990.

[76] Kyo C. Kang, Sholom G. Cohen, James A. Hess, William E. Novak and A. Spencer Peterson. *Feature-Oriented Domain Analysis (FODA) Feasibility Study*. Technical Report CMU/SEI-90-TR-021, Software Engineering Institute, Carnegie Mellon University, 1990.

[77] Ruogu Kang, Laura Dabbish, Nathaniel Fruchter and Sara Kiesler. My data just goes everywhere: User mental models of the internet and implications for privacy and security. *Symposium on Usable Privacy and Security* (SOUPS), Jul. 2015.

[78] Mitchell Kapor. A software design manifesto. Reprinted as Chapter 1 of [149].

[79] Tom Kelley and David Kelley. *Creative Confidence: Unleashing the Creative Potential Within Us All*. Crown Business, 2013.

[80] Johan de Kleer and John Seely Brown. Mental models of physical mechanisms and their acquisition. In J. R. Anderson (ed.), *Cognitive Skills and Their Acquisition*, Lawrence Erlbaum, 1981, pp. 285–309.

[81] Amy J. Ko and Yann Riche. The role of conceptual knowledge in API usability. *IEEE Symposium on Visual Languages and Human-Centered Computing* (VL/HCC), 2011, pp. 173–176.

[82] Amy J. Ko. The problem with "learnability" in human-computer interaction. *Bits and Behavior Blog*, February 16, 2019. At https://medium.com/bits-and-behavior/the-problem-with-learnability-in-human-computer-interaction-91e598aed795.

[83] Amy J. Ko, with contributions from Rachel Franz. *Design Methods*. Full text at https://faculty.washington.edu/ajko/books/design-methods.

[84] James Koppel and Daniel Jackson. Demystifying dependence. *Proceedings of the ACM SIGPLAN International Symposium on New Ideas, New Paradigms, and Reflections on Programming & Software* (Onward! 2020), 2020.

[85] Leslie Lamport. The temporal logic of actions. *ACM Transactions on Programming Languages and Systems*, Vol. 16:3, May 1994, pp. 872–923.

[86] Butler W. Lampson. *Principles of Computer Systems*, 2006. At http://www.bwlampson.site/48-POCScourse/48-POCS2006.pdf.

[87] Axel van Lamsweerde. Goal-oriented requirements engineering: A guided tour. *Fifth IEEE International Symposium on Requirements Engineering* (RE'01), 2001.

[88] Axel van Lamsweerde and Emmnuel Letier. Handling obstacles in goal-oriented requirements engineering. *IEEE Transactions on Software Engineering*, Vol. 26:10, Oct. 2000, pp. 978–1005.

[89] Axel van Lamsweerde. *Requirements Engineering: From System Goals to UML Models to Software Specifications*. Wiley, 2009.

[90] Bruno Latour. Where are the missing masses? The sociology of a few mundane artifacts. In Wiebe Bijker and John Law (eds.), *Shaping Technology/Building Society: Studies in Sociotechnical Change*, MIT Press, 1992, pp. 225–258.

[91] Michael Leggett. The evolution of Gmail labels. July 1, 2009. At https://googleblog.blogspot.com/2009/07/evolution-of-gmail-labels.html.

[92] Nancy G. Leveson and Clark S. Turner. An investigation of the Therac-25 accidents. *Computer*, Vol. 26:7, July 1993, pp. 18–41.

[93] Matthys Levy and Mario Salvadori. *Why Buildings Fall Down: How Structures Fail*. Norton, 1992.

[94] Karl J. Lieberherr and Ian Holland. Assuring good style for object-oriented programs. *IEEE Software*. Vol. 6:5, Sept. 1989, pp. 38–48.

[95] Barbara Liskov and Stephen Zilles. Programming with abstract data types. *Proceedings of the ACM SIGPLAN Symposium on Very High Level Languages*, 1974, pp. 50–59.

[96] Barbara Liskov and John Guttag. *Abstraction and Specification in Program Development*. MIT Press, 1986.

[97] Vernon Loeb. 'Friendly fire' deaths traced to dead battery. *Washington Post*, March 24, 2002.

[98] Donna Malayeri and Jonathan Aldrich. Is structural subtyping useful? An empirical study. *Proceedings of the European Symposium on Programming* (ESOP '09), March 2009.

[99] George Mathew, Amritanshu Agrawal, and Tim Menzies. Trends & topics in software engineering. Presentation at Community Engagement Session, *International Conference on Software Engineering* (ICSE'17), 2017. At http:// tiny.cc/tim17icse.

[100] Steve McConnell. *Code Complete: A Practical Handbook of Software Construction*, 2nd Edition. Microsoft Press, 2004.

[101] Malcolm Douglas McIlroy. Mass produced software components. *Software Engineering: Report of a Conference Sponsored by the NATO Science Committee*, Garmisch, Germany, 7–11 Oct. 1968.

[102] George H. Mealy. Another look at data. *Proceedings of the Fall Joint Computer Conference* (AFIPS '67), 1967, pp. 525–534.

[103] Thomas P. Moran. The command language grammar: A representation for the user interface of interactive computer systems? *International Journal of Man-Machine Studies*, Vol. 15:1, Jul. 1981, pp. 3–50.

[104] Steven J. Murdoch, Saar Drimer, Ross Anderson and Mike Bond. Chip and PIN is Broken. *31st IEEE Symposium on Security and Privacy* (S&P 2010), 2010.

[105] John Mylopoulos. Conceptual modeling and Telos. In P. Loucopoulos and R. Zicari (eds.), *Conceptual Modelling, Databases and CASE: An Integrated View of Information Systems Development*, McGraw Hill, New York, 1992.

[106] Jakob Nielsen. Usability engineering at a discount. *3rd International Conference on Human-Computer Interaction*, Sept. 1989.

[107] Jakob Nielsen and Rolf Molich. Heuristic evaluation of user interfaces. *Proceedings of the SIGCHI Conference on Human Factors in Computing Systems* (CHI '90), 1990.

[108] Jakob Nielsen. 10 Usability Heuristics for User Interface Design. 1994. At https://www.nngroup.com/articles/ten-usability-heuristics.

[109] Robert L. Nord. *Deriving and Manipulating Module Interfaces*. Doctoral Dissertation, School of Computer Science, Carnegie Mellon University, 1992.

[110] Donald Norman. *The Design of Everyday Things*. Originally published under the title *The Psychology of Everyday Things*. Basic Books, 1988.

[111] Donald Norman. *The Design of Everyday Things: Revised and Expanded Edition*. Basic Books, 2013.

[112] Donald Norman and Bruce Tognazzini. How Apple is giving design a bad name. *Codesign*, November 10, 2015. At http://www.fastcodesign.com/3053406/how-apple-is-giving-design-a-bad-name.

[113] Omnicognate Blog. In CSS, "px" is not an angular measurement and it is not non-linear. January 7, 2013. At https://omnicognate.wordpress.com/2013/01/07/in-css-px-is-not-an-angular-measurement-and-it-is-not-non-linear.

[114] Shira Ovide. No, the best doesn't win. *New York Times*, April 27, 2020. At https://www.nytimes.com/2020/04/27/technology/no-the-best-doesnt-win.html.

[115] David L. Parnas. On the criteria to be used in decomposing systems into modules. *Communications of the ACM*, Vol. 15:12, Dec. 1972, pp. 1053–1058.

[116] David L. Parnas. Designing software for ease of extension and contraction. *IEEE Transactions on Software Engineering*, Vol. 5:2, March 1979.

[117] Chris Partridge, Cesar Gonzalez-Perez and Brian Henderson-Sellers. Are conceptual models concept models? *Proceedings of the 32nd International Conference on Conceptual Modeling* (ER 2013), Volume 8217, Springer-Verlag, 2013.

[118] Santiago Perez De Rosso and Daniel Jackson. Purposes, concepts, misfits, and a redesign of git. *ACM SIGPLAN International Conference on Object-Oriented Programming, Systems, Languages, and Applications* (OOPSLA 2016), 2016, pp. 292–310.

[119] S. Perez De Rosso, D. Jackson, M. Archie, C. Lao, and B. McNamara III. Declarative assembly of web applications from predefined concepts. *Proceedings of the ACM SIGPLAN International Symposium on New Ideas, New Paradigms, and Reflections on Programming & Software* (Onward! 2019), 2019.

[120] Charles Perrow. *Normal Accidents: Living with High-Risk Technologies*. Princeton University Press. Revised edition, 1999.

[121] Henry Petroski. *To Engineer Is Human: The Role of Failure in Successful Design*. Vintage Books, 1985.

[122] Malte Plath and Mark Ryan. Feature integration using a feature construct. *Science of Computer Programming*, 41(1):53–84, 2001.

[123] Amir Pnueli. The temporal logic of programs. *Proceedings of the 18th Annual Symposium on Foundations of Computer Science* (FOCS), Nov. 1977.

[124] Michael Polanyi. *The Tacit Dimension*, University of Chicago Press, 1966.

[125] Michael Polanyi. *Personal Knowledge: Towards a Post-Critical Philosophy*. University of Chicago Press, 1974.

[126] Edouard de Pomiane. *French Cooking in Ten Minutes: Adapting to the Rhythm of Modern Life*. Original, 1930; English translation, Farrar, Strauss and Giroux, 1977.

[127] Colin Potts. Using schematic scenarios to understand user needs. *ACM Symposium on Designing interactive Systems: Processes, Practices and Techniques* (DIS'95), Aug. 1995.

[128] Trygve Reenskaug with Per Wold and Odd Arild Lehne. *Working with Objects: The OOram Software Engineering Method*. Manning/Prentice Hall, 1996.

[129] Kerry Rodden and Michael Leggett. Best of both worlds: Improving gmail labels with the affordances of folders. *ACM Conference on Human Factors in Computing Systems* (CHI 2010), Apr. 2010.

[130] David Rose. *Enchanted Objects*. Scribner, 2014.

[131] James Rumbaugh, Michael Blaha, William Premerlani, Frederick Eddy and William Lorensen. *Object-Oriented Modeling and Design*. Prentice Hall, 1994.

[132] Jerome H. Saltzer and M. Frans Kaashoek. *Principles of Computer System Design: An Introduction*. Morgan Kaufmann, 2009.

[133] Mario Salvadori. *Why Buildings Stand Up: The Strength of Architecture*. Norton, 1980.

[134] Ben Shneiderman, Catherine Plaisant, Maxine Cohen, Steven Jacobs and Niklas Elmqvist. *Designing the User Interface: Strategies for Effective Human-Computer Interaction*, Sixth Edition, Pearson, 2016.

[135] Herbert A. Simon. The architecture of complexity. *Proceedings of the American Philosophical Society*, Vol. 106:6, Dec. 1962, pp. 467–482.

[136] John Michael Spivey. *The Z Notation: A Reference Manual*. International Series in Computer Science (2nd ed.), Prentice Hall, 1992. Full text at https://spivey.oriel.ox.ac.uk/wiki/files/zrm/zrm.pdf.

[137] Nam P. Suh. *The Principles of Design*. Oxford Series on Advanced Manufacturing (Book 6), Oxford University Press, 1990.

[138] Alfred Tarski and John Corcoran (ed.). What are logical notions? *History and Philosophy of Logic*, 7:143–154, 1986.

[139] Harold Thimbleby. *User Interface Design*, ACM Press, 1980.

[140] Harold Thimbleby, Jeremy Gow and Paul Cairns. Misleading behaviour in interactive systems. *Proceedings of the British Computer Society HCI Conference*, Research Press International, 2004.

[141] Harold Thimbleby. *Press On: Principles of Interaction Programming*, MIT Press, 2007.

[142] Harold Thimbleby. *Fix IT: See and Solve the Problems of Digital Healthcare*. Oxford University Press, 2021.

[143] Bruce Tognazzini. *First Principles of Interaction Design*, revised & expanded, 2014. At https://asktog.com/atc/principles-of-interaction-design.

[144] Jan Tschichold. *The Form of the Book: Essays on the Morality of Good Design*. Hartley & Marks, 1975.

[145] Karl Ulrich. *Computation and Pre-parametric Design*. Doctoral Dissertation, Department of Mechanical Engineering, MIT, July 1988.

[146] Bret Victor. *A Brief Rant on the Future of Interaction Design*, November 8, 2011. At http://worrydream.com/ABriefRantOnTheFutureOfInteractionDesign.

[147] Walter G. Vincenti. *What Engineers Know and How They Know It*. Johns Hopkins University Press, 1990.

[148] Forrest Wickman. What was "poking"? Maybe even Mark Zuckerberg doesn't know. *Slate*, February 4, 2014. At https://slate.com/technology/2014/02/facebooks-poke-function-still-a-mystery-on-the-social-networks-10th-anniversary.html.

[149] Terry Winograd with John Bennett, Laura De Young, and Bradley Hartfield (eds.), *Bringing Design to Software*. Addison-Wesley, 1996.

[150] Gary Wolf. Steve Jobs: The next insanely great thing. *Wired Magazine*, February 1, 1996. At https://www.wired.com/1996/02/jobs-2.

[151] Edward Yourdon. *Modern Structured Analysis*. Prentice Hall, 1989. Chapter 10: Data Dictionaries.

[152] Eric S. K. Yu. Towards modeling and reasoning support for early-phase requirements engineering. *Proceedings of the 3rd IEEE International Symposium on Requirements Engineering* (RE '97), 1997.

[153] Pamela Zave and Michael Jackson. Conjunction as composition. ACM Transactions on Software Engineering and Methodology, Vol. 2:4, 1993.

[154] Pamela Zave. Secrets of call forwarding: A specification case study. In *Formal Techniques for Networked and Distributed Systems (FORTE)*, 1995.

[155] Pamela Zave and Michael Jackson. Four dark corners of requirements engineering. *ACM Transactions on Software Engineering and Methodology*, Vol. 6:1, Jan. 1997, pp. 1–30.

Index of Applications

Index of Concepts

Index of Names

Index of Topics